CALIFORNIA NATURAL HISTORY GUIDES

INTRODUCTION TO
CALIFORNIA CHAPARRAL

California Natural History Guides

Phyllis M. Faber and Bruce M. Pavlik, General Editors

Introduction to

CALIFORNIA CHAPARRAL

Ronald D. Quinn
Sterling C. Keeley

Illustrations by Marianne D. Wallace

UNIVERSITY OF CALIFORNIA PRESS
Berkeley · Los Angeles · London

To our families, without whose love and support none of this would have been possible.

University of California Press, one of the most distinguished university presses in the United States, enriches lives around the world by advancing scholarship in the humanities, social sciences, and natural sciences. Its activities are supported by the UC Press Foundation and by philanthropic contributions from individuals and institutions. For more information, visit www.ucpress.edu.

California Natural History Guide Series No. 90

University of California Press
Berkeley and Los Angeles, California

University of California Press, Ltd.
London, England

Library of Congress Cataloging-in-Publication Data

Quinn, Ronald D.
    Introduction to California chaparral / Ronald D. Quinn, Sterling C. Keeley ; with line drawings by Marianne D. Wallace.
        p.   cm. — (California natural history guides ; 90)
    Includes bibliographical references and index.
    ISBN-13 978-0-520-21973-1 (cloth : alk. paper), ISBN-10 0-520-21973-2 (cloth : alk. paper)
    ISBN-13 978-0-520-24566-2 (pbk. : alk. paper), ISBN-10 0-520-24566-0 (pbk. : alk. paper)
1. Chaparral ecology—California.   2. Chaparral—California.   I. Keeley, Sterling C. II. Title. III. Series.

QH105.C2Q56   2006
577.3'8'09794—dc22   2004030027

Manufactured in China
10   09   08   07   06
10   9   8   7   6   5   4   3   2   1

The paper used in this publication meets the minimum requirements of ANSI/NISO Z39.48-1992 (R 1997) (Permanence of Paper). ♾

Cover photograph: Chamise chaparral mixed with manzanitas, oaks, and pines, near the northern limit of chamise, Beegum Gorge, Shasta County. Photograph and plant identification by John Sawyer.

The publisher gratefully acknowledges the generous
contributions to this book provided by

the Gordon and Betty Moore Fund
in Environmental Studies
and
the General Endowment Fund of the
University of California Press Foundation.

CONTENTS

PREFACE

The purpose of this book is to provide basic information about the chaparral community, an important but often poorly understood part of the natural heritage of California. We want others to come to appreciate the beauty, complexity, and resiliency of this quintessentially Californian ecological community. We hope that this book will convey some of the excitement and interest we have enjoyed as explorers in this intricate, fascinating environment. The plants and animals that thrive in this system, with its recurring "natural disasters," can teach us how to understand this part of California's varied landscape. This understanding encompasses the plants and animals that make up the chaparral community, the climate that shaped it, and the role of fire, drought, and floods that are recurring parts of this system. We also hope to supply insight into the growing conflict between the dynamic interaction of natural processes and urbanization as cities press up against wild chaparral landscapes, and to provide suggestions for productive courses of action that could mitigate this conflict.

The book treats the physical features of chaparral environments first, then examines common species of plants and animals more closely, and finally addresses the interactions between chaparral and humans. We have written about the plants, animals, and natural processes that people are likely to notice. Chaparral is a subject of great interest to natural scientists, and a source of fascination to specialists and generalists alike. We hope that readers of this book will find chaparral sufficiently interesting to continue to make discoveries for themselves. The idea of a concise, general book about California chaparral came to each of the authors independently. We were both concerned that there was no such book to explain the chaparral community either to students or other interested people. We each needed such a book for college teaching, and also to explain to friends and family about the system we find so fascinating to study. An earlier book of this kind, *The Elfin-Forest*, by Francis M. Fultz, is a delightful, if dated, account of the subject, which was last published in 1927 and has long since been out of print. Both of us wrote a manuscript for a chaparral book without knowledge of the activities of the other. Two years ago the University of California Press informed us of our parallel efforts, and we decided to combine forces.

Both of us have worked in chaparral as research scientists and educators for more than 30 years, and both of us have often been asked, "What do you do out there in that brush anyway?" We have set out here to convey the fascination that has drawn us back again and again to chaparral. As college instructors we have used chaparral as a classroom subject for studying ecology and land management. Sterling Keeley has a primary background in plants, and Ronald Quinn in animals, but both of our interests have always been centered on ecological processes in chaparral, especially with respect to fire.

The order in which authors are listed is the result of a coin toss and implies nothing about relative importance of contributions. We have written the entire book together.

A book of this scope encompasses the ideas and work of many people who have devoted their lives to understanding and managing chaparral. The format of this volume does not allow us to recognize most of these people by name in the text. We have acted as reporters of their accomplishments and gratefully acknowledge their work, published and otherwise, which has permitted us to produce this book.

ACKNOWLEDGMENTS

I give heartfelt thanks to my spouse, Barbara Ellis-Quinn, who in countless ways has acted as a patient and supportive midwife for the rather protracted birth of this book. She reviewed the manuscript in its entirety and created a quiet office and a warm home for writing. My friend Ken Montgomery, a botanist and horticulturist with a true love of plants native to California, provided hours of discussion and dreaming when I first thought to write this book a half lifetime ago. He has given me support, encouragement, and suggestions when they were most needed. The Rancho Santa Ana Botanic Garden of Claremont, California, has provided many days of pleasure and inspiration, as well as a place to take photographs for this book. Steve Shirley and his airplane gave me a new perspective and some photographs of wildfires at the urban-wildland interface. Rick Halsey's boundless enthusiasm was inspirational, as was his generosity with his ideas. Mike Raugh, mathematician and accomplished naturalist, reviewed a large part of the manuscript. Colleagues Chris Brady, Glenn Stewart, and Gary Wallace reviewed various parts of the manuscript. William D. Wagner provided insights on birds. Jim Dole, who first introduced me to the delights and challenges of chaparral ecology, set me on a path that has lasted a lifetime. Finally, I am grateful to students of chaparral biology at California State Polytechnic University, Pomona, who over the years have been an endless source of enthusiasm to me for all things to do with chaparral.

Ronald D. Quinn
February 2006

This book would never have come into being were it not for the inspiration and support of many people. For starting my career in chaparral research and encouraging my efforts along the way, Harold Mooney of Stanford University, deserves my deepest gratitude. Nancy Coile, of the University of Georgia, provided the initial inspiration to write this book while visiting and collecting ceanothus. My sister, Christine Thompson, willingly took hikes with me while I photographed and pontificated, read countless drafts, and buoyed up my enthusiasm when it flagged. Frank Hovore spent many hours providing interesting information about insects and animals of the chaparral, and reading drafts, as well as providing some fabulous photos. Steve Davis, Robert Gustafson, Linda Hardie-Scott, Charles Hogue, Dave Minor, Philip Rundel, Timothy Thomas, Sherry Wood, and Paul Zedler provided wonderful photographs and have been waiting for them to be returned since the early 1990s. Their generosity and patience are much appreciated. Shirleen Gudmuson, Christa Hatch, Linda Kate Schroeder, and Peter Vroom all contributed to editing, reviewing, and refining the manuscript during its various incarnations. My students at Whittier College allowed me to share my delight in the chaparral and to do original research that would not have been possible without them. They taught me to see the natural world in new ways, for which I am truly grateful. Thanks to Phil Rundel and one anonymous reviewer for contributing to the quality of the book. And finally, our thanks to the editors, Phyllis Faber, who suggested we work together on this project, and Doris Kretschmer, who worked out the details and kept us on track.

Sterling C. Keeley
February 2006

CHAPARRAL IS BOTH a vegetation type and the name given to the community of coadapted plants and animals found in the foothills and mountains throughout California. The chaparral vegetation is composed of a diverse assemblage of different species of evergreen drought- and fire-hardy shrubs. Seen from the car window or scenic lookout, chaparral looks like a soft bluish green blanket gently covering the hills. Up close, however, this "blanket" no longer appears soft. Instead, what is revealed is a nearly impenetrable thicket of shrubs with intertwined branches and twigs with hard leaves and stiff and unyielding stems. The shrubs are well adapted to the rigors of long, hot, dry summers and unpredictable winter rainfall that are characteristic of California's mediterranean climate. Chaparral is especially extensive in the central and southern parts of the state, but it covers large areas of northern California as well (pl. 1).

A stand of mature chaparral (10 years old or more) is usually composed of shrubs of the same age and approximate

Plate 1. Mountain chaparral *(foreground)* is typically found adjacent to pine and oak forests, as seen in the distance (Mendocino County).

Plate 2. This view into the understory of a stand of hoaryleaf ceanothus in the San Gabriel Mountains shows the dense growth of shrub stems and the absence of an understory beneath mature chaparral. The metal objects in the foreground are seed traps.

height, dating from the last fire. The canopy height can range from waist level to 20 feet tall. While the shrubs may be quite tall, the leaves are found only in the upper portions where strong sunlight reaches. Below the leafy canopy the shrubs have one to several rigid, woody stems that arise from a common base (pl. 2). These stems range in size from one to three inches in diameter in young stands and up to 12 inches thick in older stands. Mature shrubs grow so close together that the branches of adjacent plants are interlaced, forming an unbroken layer of vegetation with few openings. Beneath the shrubs the dimly lit ground lacks any vegetation and is bare except for a sparse litter of dead leaves and twigs. In many species of chaparral shrub there is a burl or root crown, an enlarged woody mass at or slightly below the soil surface at the base of the stems. After fire, if the above-ground portion of the plant has been killed, the burl produces new shoots fed from a deep root system. The roots may travel 100 feet or more in search of water. Growing both horizontally and vertically, the root systems of chaparral shrubs form a matrix that holds the soil in place on the hillsides (fig. 1). Despite the underlying hard nature of the stems and leaves, the shrubs produce beautiful and fragrant blossoms, such as those of bigpod ceanothus (pl. 3).

Mature chaparral stands may persist for a century or more, the shrubs changing slowly over time. With the passage of time most shed branches and leaves, while others die completely.

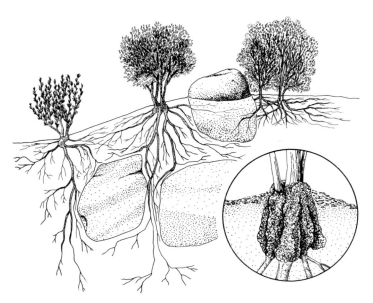

Figure 1. Root systems of the three most common types of chaparral plants: chamise *(left),* manzanita *(middle),* and ceanothus *(right)* help hold the soil. The inset shows a typical young chamise burl.

Plate 3. Bigpod ceanothus in flower, with immature fruits in the center. This species is common in southern California.

The overhead branches are quite dense in mature chaparral, but with age there is often a space beneath the shrubs that is tall enough for exploration on hands and knees. Sitting quietly beneath these shrubs, away from the sunlit hubbub above, a patient observer can take advantage of the unobstructed view near the ground to see some of the unique animal inhabitants of dense chaparral. For example, common sights are thrashers rummaging through the litter with their long curved bills, and California Whipsnakes, also known as Striped Racers, hurrying about with head held high (see chapter 5). In these natural surroundings many animals do not seem to recognize humans as dangerous and go about their business as if no one were present. Birds like the curious Wrentit will hop from branch to branch until they are almost within reach, with head cocked from side to side so that each yellow eye in turn can look closely at the strange and clumsy visitor.

The name chaparral, from *chaparro,* was given by Spanish

Plate 4. Toyon, or California holly, with fruit clusters. This is the plant that gave Hollywood its name.

colonists to refer to the place where scrub oaks grow. The California chaparral reminded them of the similar-appearing vegetation of southern Spain (see chapter 2 for more on other mediterranean climate areas). "Chaps," leather leggings worn by riders on horseback to protect against scratches from thorny vegetation, is a word that derives from the same Spanish root word.

Chaparral is built into the visual image of outdoor California even for those who have never visited the state. It forms the backdrop for thousands of movies, television productions, and videos because it is the common vegetation of the hills and mountains around Los Angeles. One chaparral plant, California holly, is, in fact, responsible for giving Hollywood its name (pl. 4). Everyone who sees filmed car chases on mountain roads, advertisements for SUVs taking on the tough hills, or the televised mountain vistas above the Rose Bowl on New Year's Day has seen chaparral.

Chaparral is a place full of life. It is an ancient and exquisitely balanced community of many kinds of plants and animals, each with its own special stories. For example, it has rain beetles that stay hidden underground for years, outlasting drought (pl. 5), fire beetles that mate only on burning branches, plant seeds that require fire to germinate, lizards that shoot blood from their eyes when threatened, kangaroo rats that never drink water, and wood rats that collect seeds, forks, tire treads, and a host of other strange objects to build up their nests. We include these and other stories throughout the book.

Plate 5. A male rain beetle preparing to fly in search of a mate.

Fire and Chaparral

Chaparral has always existed with fire. It is this natural "disaster" that often brings chaparral to public attention, but it is a normal and natural part of life in the chaparral. A cycle of recovery and new birth is initiated by the burning off of the shrubs that make up the mature chaparral. The seeds produced by many chaparral shrubs require fire to germinate. Specialized short-lived plants, called fire annuals, appear only in response to fire even if they have to wait more than 100 years between blooms! Similarly, deer, birds, lizards, and insects use the lush new growth that appears after fire for food and reproduction. Fires do not touch all chaparral areas in a single year, and some areas may be missed by fire for a century or more. This does mean that very large fires as well as many smaller ones may happen apparently randomly in different parts of the state. The irregularity of fires is natural as well. Fire frequency depends on the condition of the vegetation and the interaction of factors such as ignition source, winds, season, topography, and time elapsed since the previous fire. This unpredictability creates a patchwork of recently burned and long-unburned chaparral across the state.

Plate 6. Mariposa lilies come in many colors: pink, white, orange, red, or yellow as shown here. Many have patterns of stripes, dots, and hairs to aid in pollination (Santa Monica Mountains).

Shrubs surrender the ground to lower and softer plants for a brief period after fire. At this time some of the most beautiful wildflowers in the state appear in great numbers, forming a brilliant and colorful carpet quite different from the tangle of shrubs that preceded them. California is known for its poppies, but in addition to these, after a chaparral fire there are also whispering bells, lilies (pl. 6), snapdragons, phacelias, and dozens of species of small flowering plants found at no other time or place. Chaparral recovers quickly after fire, spreading a fresh mantle of shrubs upon the hillsides within a few years. The chaparral has existed for many thousands of years in California, and fire has always been an integral part of this community. In short, no fire, no chaparral.

Life in chaparral is unpredictable. It is a place that is fierce and unrelenting, fabulously beautiful, and prone to disaster and yet has persisted for countless millennia. The juxtaposition of the natural processes of the chaparral ecosystem with growing urban areas produces difficulties on both sides. Many of us live in intimate association with chaparral, for better or worse, so it is important that we come to understand it.

The modern era of urban-wildland chaparral fire holocausts, so prominent in news stories each fall, opened with the Bel Air fire of November 1961 when Santa Ana winds drove a fire out of the chaparral of the Hollywood Hills into an enclave populated by the rich and famous. In a few hours almost 500 of some of the most expensive houses in the state were gone. Aldous Huxley lost a lifetime collection of books and papers that might have become the pride of a research library. After that event he called himself "a man without a past." One celebrity is reputed to have remarked, "Things like this shouldn't happen in such a nice neighborhood." Indeed, they should not. But despite this sentiment, things like this do happen in all kinds of neighborhoods and will continue to do so as long as we persist in building flammable houses in fire-prone settings, such as chaparral-covered hillsides and canyons (pls. 7, 75, 78). A house perched on a ridge atop a chaparral-filled canyon is almost sure to be threatened by fire at some point, especially if it has large wooden decks, overhanging eaves, and a wooden roof. While many communities now make fire-resistant roofing a requirement and others insist on low-flammability vegetation surrounding a house, the overall effect is imperfect. The natural functioning of the chaparral includes the potential for catastrophic mudslides, as well as fire. Mudslides may result from the combination of heavy rains and steep hillsides recently denuded of chaparral by fire.

Plate 7. Chaparral wildfires can spread over large areas, as seen in this photograph from the Malibu fire in 1985.

We like to build our communities where we want, despite clear indications of danger from fire and subsequent mudslides. This is costly for all concerned (see chapter 6 for further explanation). Fire is an intrinsic part of the chaparral, as it has been for many thousands of years, and it is up to us individually and collectively to take responsible action recognizing this inevitability.

Where Is Chaparral Found?

Chaparral is never far from sight in much of California. The tourist at San Diego's animal parks looking to the east, the Los Angeles commuter idly gazing to the north, a school child in the Great Valley watching the eastern horizon for airplanes, and the Bay Area resident on the way to Tahoe and scanning the Sierra for signs of snow will all see hillsides covered with chaparral. Chaparral provides the shrubby covering of the foothills ringing the populated valleys and coastal plains of the state. Many suburban dwellers now live surrounded by chaparral.

Chaparral covers approximately 7 million acres of California. It is found on coastal and inland mountain slopes throughout the state west of the deserts, north into southwestern Oregon, and south into Baja California (map 1).

Chaparral is most extensive and diverse from the Central Coast Ranges south and inland to the interior edges of the South Coast, Transverse, and Peninsular Ranges, and south of the international border to the southern end of the Sierra San Pedro Martir (map 1). Chaparral is found along coastal bluffs and mountains, around the fringes of valleys, up foothills, and across entire mountain ranges. Vast and continuous tracts of chaparral cover most of the interior of the counties of Monterey, San Luis Obispo, Santa Barbara, Ventura, Orange, and San Diego. In these places chaparral is often the dominant vegetation as far as the eye can see (pl. 8). In northern California, chaparral blankets the hillsides of the San Francisco Peninsula and grows thickly in Lake and Mendocino Counties and inland farther north. At points it may descend to the coast, and chaparral intergrades with forests as it moves north. Chaparral also covers much of the Sierran foothills, as can be clearly seen on the drive from the west into Yosemite National Park.

Chaparral Is Found with Other Vegetation Types

California's landscape is heterogeneous, and as a result many different local environmental conditions exist side by side. This gives rise to much variation in plant and animal distributions over short distances. Latitude and elevation repro-

Map 1. Major topographical features and the general distribution of chaparral in California. The shaded areas indicate chaparral. These areas exaggerate the total area occupied by chaparral because in some places they include associated plant communities such as oak and pine woodlands, and coastal sage scrub.

Plate 8. Parks throughout the state, like this one in the Santa Monica Mountains in southern California, offer hiking and other recreational activities in chaparral.

duce many of the same temperature and rainfall gradients; for example, a high mountain in southern California may be as moist and cool as a low-elevation area farther north. Local topographic and soil factors also affect the environment. Sometimes this means that vegetation types such as chaparral (pls. 9, 10) will be juxtaposed or intermingle with elements of pine or oak forests or border grasslands, or mingle with desert or coastal species.

Because of this physical heterogeneity, chaparral, which is widespread, does vary in composition from place to place. The plant species present in each area are those best suited to the particular local climatic conditions, soils, and topography, and their distributions are independent of those of other species. Chaparral shrubs may be found as occasional members of other communities as well. The boundaries between chaparral and other communities may be distinct, as where pro-

Plate 9. Flowering chaparral yuccas, a common sight in chaparral, particularly after fire. Scarlet bugler and a white-flowered ceanothus shrub appear in the foreground (Lone Pine Canyon, San Gabriel Mountains, May 1971).

Plate 10. Geological movements have produced rocky exposures and mixed soil types on which chaparral grows, along with trees and grasses. Oaks forest covers the north-facing slope of the mountain (left of the rocky ridge) and dots the lower slopes, while grasses make up the light green areas and chaparral forms the solid medium green cover over most of the hillside (Mount Diablo, Contra Costa County).

Plate 11. Chamise covers the lower south-facing slopes (dark green), while manzanita dominates the ridges and higher slopes (light green) in the central Sierra.

truding rock outcrops and steep ridges separate them (pl. 10), or indistinct, as where conditions favor a mixture of species. Thus, naming chaparral or other vegetation types is somewhat arbitrary and can only be an approximation of natural patterns. We create names for convenience in referring to different vegetation types.

Overall the chaparral is similar enough throughout the state that it allows us to recognize this vegetation and community type wherever it is found. However, the differences from region to region have been recognized and given names for geographic areas, for example, North Coast, Sierran, or desert chaparral; for distinctive soil types, as in serpentine chaparral; or for the most common shrub species. In this latter case, manzanita chaparral (pl. 11), ceanothus chaparral, scrub oak chaparral, and chamise chaparral are the most commonly applied names. These various subclassifications of chaparral are useful at the local level to land managers and others in the community who deal with particular areas on a day-to-day basis. At least 48 such chaparral classifications have been proposed and are used by state agencies and conservation organizations (see *A Manual of California Vegetation*, in the supplemental readings section).

Coastal Sage Scrub Is Not Chaparral

Coastal sage scrub is a low-growing drought-deciduous shrubby vegetation type found in the southern half of California. It is sometimes confused with chaparral and can grow near it, but it is a distinct vegetation type in its own right. Probably the most characteristic aspect of coastal sage vegetation is its smell, a pungent, spicy aroma that carries for great distances. Although both vegetation types are shrubby, a notable difference between chaparral and coastal sage is that while chaparral plants are evergreen, coastal sage plants are drought-deciduous, losing their leaves during the hot part of the year. Coastal sage shrubs are also typically shorter (three to six feet tall), with pale, soft leaves and stems, and there are gaps between shrubs. The foliage is pale in color due to a covering of white or gray hairs on one or both sides of the leaves. The dominant shrubs of this vegetation type are the sages, buckwheats, and California sagebrush. Coastal sage plants can exist in drier areas than can the evergreen chaparral shrubs and so are also found on dry hilltops, ridges, and outcrops in chaparral communities in southern California. They are also characteristic of disturbed areas along roadsides, sometimes far inland. Because plants have individual preferences and tolerances, occasionally a chaparral shrub such as lemonadeberry or laurel sumac will be found in areas dominated by coastal sage and vice versa.

How Organisms Are Named

Throughout this text plants and animals are identified by their common name and their scientific name (fig. 2). A common name, such as "deerweed" or "Desert Cottontail," is like a person's nickname. It is good enough for most purposes when used by people who are familiar with that particular species in that particular geographical area. Since common names can vary and are not uniformly applied everywhere, each species of plant and animal also has an official scientific name, a genus and species that identify it uniquely. The unique genus and species name for human beings, for example, is *Homo sapiens.*

Figure 2. The Desert Cottontail beside deerweed.

For deerweed the scientific name is *Lotus scoparius,* and for Desert Cottontail it is *Sylvilagus audubonii.* Genus and species names are given in parentheses after the common name of an organism throughout the text, usually the first time it is mentioned. The full scientific name sometimes includes a third name, which designates a subspecies. Using one of the names above, this might be *Lotus scoparius* subspecies *scoparius,* for example. This trinomial identifies a particular group of organisms as being members of a distinct subspecies confined to a particular geographical area and having distinctive morphological features. Scientific names are typically given in italics because they are in Latin or are latinized and therefore are not English words.

Additionally, the genus and species names fit within other larger categories such as families and orders that express wider relationships to other groups of plants or animals. For example, both rabbits and mice (and indeed humans) are part of an even larger group of animals in the class Mammalia. Other large classes of animals are birds (Aves), reptiles (Reptilia), and insects (Insecta). Rabbits belong to the animal order Lagomorpha, which distinguishes it from other groups of small ground mammals such as mice, which belong to the order Rodentia (see chapter 5). Similar kinds of plants and animals are grouped under their order or family in the text, depending on the most common convention of taxonomists for each group. There are differences between the commonly used level of classification for plants and for animals, as you will discover. For example, for animals and insects the level of order is commonly used to group similar types together, whereas in flowering plants the level of family is more typical. For the scientific naming of plants and animals we have used the most up-to-date references at the time of writing. For example, *The Jepson Manual: Higher Plants of California* (see the supplemental readings section) was used to standardize plant names. As scientists study relationships between and within species by using new information, such as that from DNA sequencing, the names of some organisms are changed to reflect this new understanding. Therefore, with the passage of time the reader may encounter slight differences in scientific names between the ones used in this book and those in other publications.

CALIFORNIA CHAPARRAL IS a product of the mediterranean climate in which it grows, where summers are hot and dry and winters mild and wet. Less than three percent of the world's land surface has a mediterranean climate. This climate type is found in California, the Mediterranean Basin, central Chile, the southern tip of Africa, and parts of Australia (map 2). Each of these areas has an evergreen shrubby vegetation commonly referred to as chaparral, maquis, garrigue, matorral, fynbos, or heath. The environmental conditions that produce a mediterranean climate result from the confluence of three factors: latitude, a cold ocean, and a large high-pressure air mass that extends from the ocean across the land of the adjacent continent. These factors come together on the western edge of continents between roughly 30 and 40 degrees latitude on either side of the equator. The high-pressure air mass shifting north and south with the seasons produces the alternating cool, wet winters and hot, dry summers characteristic of this climate type.

The Pacific High

The Pacific High air mass controls California's mediterranean climate. This slowly circling and gradually descending mass of air produces high pressure from the central Pacific Ocean all the way to California. Areas beneath this air mass are largely protected from storms half the year, producing rainless summers and falls. From midspring until late fall the Pacific High covers a large portion of California's coast and nearby inland areas. In winter, the sun moves to the south and the Pacific High travels with it. As the Pacific High drifts south, storm systems from the North Pacific can move progressively farther down the California coast, eventually reaching as far south as Baja California. Other storms from the east, over the Great Plains, for example, can also travel west to California because the Pacific High no longer deflects them. Be-

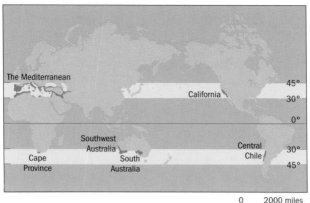

Map 2. Mediterranean climate is found in five locations worldwide between 30 and 45 degrees latitude as shown in red. (Scale reflects distances at equator.)

cause of the immense barrier of the Pacific High, the driest time of the year in California is typically in June, July, and August (fig. 3). The Pacific high exerts its influence over southern California for a longer period of the year than over northern California. Consequently, the rainy season is somewhat longer and more intense in the north.

Rainfall—Always Unpredictable

The amount of rainfall varies greatly from winter to winter. This variability occurs over the entire state, so even though northern California is wetter overall than southern California (fig. 3), in some years areas in the north may be as dry or drier than areas much farther to the south. Figure 4 shows rainfall measurements taken at the Los Angeles Civic Center over a

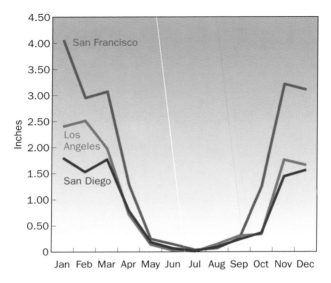

Figure 3. Average monthly rainfall in San Francisco, Los Angeles, and San Diego from 1961 to 1990. Northern California has a slightly shorter period of summer drought than southern California and receives more rainfall during the rainy season. (Data from Department of Meteorology, University of Utah.)

20-year period between 1982 and 2002. The average rainfall over this period was approximately the same as over the past century: 15 inches per rainfall year (July 1 to June 30). Twelve-month totals are measured from the middle of the calendar year so that the November through April rainy season is totaled together. This makes ecological sense as well, since most chaparral plants grow between late winter and early summer in response to the preceding rainy season. It is not only the multiyear average rainfall that matters, but also the range of variation from year to year. This can be large in California. For example, in the 2001/02 rainfall year, the total was less than one-third of the average amount, while the 1997/98 rainfall year had more than twice the average amount. Notice that the

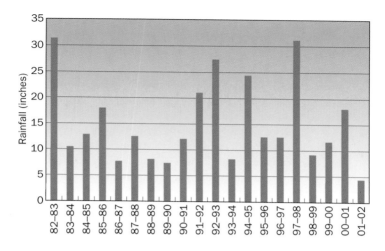

Figure 4. The annual seasonal rainfall (July 1–June 30) at the Los Angeles Civic Center over 20 years.

total fluctuates widely among years, and that the amount in any given year is not a predictor of how wet or dry the next year will be.

The spacing of storms, their duration, and their intensity are also highly variable and unpredictable. One storm may provide one-third to one-half of the total for a rainfall year all at once, while another may drop only enough moisture to dampen the surface. For example, in April 1926 a world record was set in the San Gabriel Mountains of southern California when almost an inch of rain fell in one minute. After this deluge, the region received no rain at all for several weeks. In January 1943 another record was set, also in the San Gabriel Mountains, when 26.12 inches of rain fell during a 24-hour period. Read about the impact on people of these intense storms in chapter 6.

Each rainfall year, many weeks and sometimes months have no detectable precipitation. There are occasionally even

12 consecutive months with no rain. Drought is an annual summer event in the chaparral, but it is also unpredictable in its duration and severity. A study of rainfall patterns in central California showed large variations during the winter months over a 20-year period. During that time, 12 winter months had no rainfall at all. Droughts of several years' duration are not uncommon, as exemplified between the years 1986 and 1991. These fluctuations between wet and dry are a recurrent feature of California's mediterranean climate and are often linked with cyclical variations in winds and ocean currents in the equatorial Pacific Ocean. These equatorial cycles affect weather patterns in many parts of the world. In fact, the idea of a "normal" year based on average rainfall over many years is misleading, because it fails to take into account year-to-year variability. Those who have experienced water rationing during droughts, or floods and mudslides in wet years can testify to the side effects of these wide climatic variations.

Winds That Carry Water or Take It Away

Wind is also a basic component of California's mediterranean climate and, like rain, can be either life sustaining or life threatening. The prevailing wind pattern that provides moisture in winter and cool sea breezes in summer moves from the ocean toward the mountains and inland valleys. These winds are called the Westerlies. Periodically winds flow from the east, reversing the normal wind direction. These latter winds have been variously called Santa Anas, Diablo or Devil winds, and Sundowners.

The Westerlies

The Westerlies flow regularly from the ocean onto the land all along California's coast during all seasons. These winds carry moisture inland, where it often precipitates as fog drip, rain, or dew when it reaches the foothills and mountains. The western slopes of most mountain ranges in the state are greener and the shrubs taller and thicker than on the east side of the mountains because the preponderance of precipitation brought there by the regular flow of the Westerlies is deposited on the windward side. Westerlies also moderate air temperatures, an important ingredient of California's mediterranean climate.

The winds from the ocean are not always gentle, however. During winter storms the Westerlies may carry so much water that when they reach the foothills the downpours described above can cause flash floods. If the chaparral on steep slopes has been recently burned off, the force of the rain can wash large quantities of soil and debris off the mountainsides and flush out canyons into the valleys below. The fire, besides removing the standing vegetation, often burns up the organic matter in the soil as well, causing it to become powdery and unstable. This combination of factors can result in a slug of thick mud and debris that can surge out of the canyons, over barriers, and out onto the valley floors. The force of this mud is sufficient to toss cars off roads, and crush and bury houses, leaving little evidence of former human habitation in its wake (see chapter 6).

Fire-Fanning Winds from the East

Strong winds from the east periodically change the atmosphere over chaparral completely. These are the winds responsible for fanning wildfires into raging and unstoppable blazes. They are unlike the prevailing westerly winds in every respect, whistling down mountain passes from the east, blowing hot, dry, gusty breaths through the summer-parched vegetation. Classified as foehn winds by meteorologists, they are generated by interior high-pressure air masses that are forced down mountain slopes and valleys (fig. 5). They are called Santa Ana winds in southern California, reputedly because they

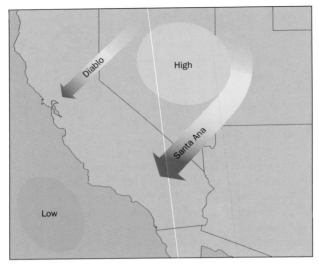

Figure 5. Santa Ana (south) and Diablo (north) winds blow strongly from high-pressure areas over the Great Basin toward the coast of California.

blow strongly through the gap in the Santa Ana Mountains and out across coastal Orange County. They are known as Diablo or Devil winds in northern California because they originate near Mount Diablo. In the Central Coast Ranges near the city of Santa Barbara, winds called Sundowners develop in the hours immediately after sunset as hot air suddenly pours down over the mountains from the nearby Santa Ynez valley. While producing a similar-feeling east wind, a Sundowner is not technically a foehn wind. In the Bay Area the Diablo winds, like the Santa Anas of the south, can drive large wildfires. Two of the most destructive wildfires in the history of the United States were driven west by downslope winds out of the vegetation of the Berkeley Hills and into the East Bay cities of Oakland and Berkeley (see chapter 6).

Surprisingly, in California these very dry winds are often

connected with storm systems. Santa Ana–type winds blow most often between September and December; however, they can occur any time of the year except summer. As the stable high pressure that anchors the long California summer begins to drift southward in fall, it allows a cavalcade of low-pressure storm systems from the Bering Sea to move across the state. In early fall, as the southern fringe of this storm track begins traveling through California, the trailing edges of storm fronts lightly brush the state but are too weak to produce rain. These weather fronts are often closely followed by high pressure that clears the sky while moving in a southeasterly direction toward the center of the continent. As these high-pressure areas drift over the Great Basin, they push a westerly flow of air back toward the Pacific Ocean (fig. 5). This river of air accelerates as it flows and tumbles down California mountain slopes. The steep drop in elevation (approximately a mile) from the Great Basin to the coast compresses, heats, and dries the air. As the winds cross the mountains, the canyons and passes become wind tunnels, constricting the flow of air through narrow passages and increasing its velocity. The interplay of rugged topography and the shifting of the high-pressure center causes sudden changes in wind direction and velocity. In concert, these physical variables produce hot, dry, erratic winds that blow across slopes clothed in chaparral and down chaparral-filled canyons.

Extreme fire weather is not confined to fall, even though this period is normally regarded as the chaparral wildfire season. The catastrophic episodes of widespread chaparral wildfires that visit California with distressing frequency are almost always whipped up by the easterly winds. They can drive choking clouds of smoke across populated valleys and far out to sea, as shown by photographs taken from space (pl. 12). Wind-driven chaparral wildfires have been described from the time Europeans first arrived in California. In December 1793, from his ship, the English explorer George Vancouver saw the following, just south of San Diego:

Plate 12. Smoke and ash from chaparral wildfires blown out to sea by Santa Ana winds, October 2003. The largest cloud *(right)* is from San Diego County, the smaller cloud *(center)* is from the San Gabriel and San Bernardino Mountains of Los Angeles and San Bernardino Counties, and a third *(upper left)* is from Ventura County. There are much smaller plumes visible from Baja California to the south, and San Luis Obispo County to the north.

During the forenoon immense columns of smoke were seen to arise from the shore in different parts....These clouds of smoke containing ashes and dust soon enveloped the whole coast....The easterly wind still prevailing, brought with it from the shore vast volumes of this noxious matter, not only uncomfortable to our feelings, but adverse to our pursuit, as it intirely [sic] hid from our view every object at the distance of an hundred yards...by the prevalence and strength of the north-east and easterly wind, spread to a very great extent. Large columns of smoke were still seen rising [the following day] from the vallies [sic] behind the hills, and extending to the northward along the coast....Under these circumstances it cannot be a matter of surprize that the country should present a desolate and melancholy appearance. (Lamb 1984, 1115–1116)

Sometimes rain does not come until well into winter, prolonging the fire season past the end of the calendar year. There have been large chaparral wildfires in California every month of the year. In March 1964, at the time when moisture in the soil and in the plants would normally make chaparral shrubs virtually fireproof, simultaneous chaparral wildfires driven by Santa Ana winds in three southern California mountain ranges burned a total of 11,000 acres.

Temperature

A mediterranean climate may feature moderate or extreme temperatures any time of the year. The most moderate climate prevails near the ocean where there may be only a 10 degree F fluctuation in average temperature over the entire year. This is one reason why many people prefer to live along the coast. Westerlies carry air from the ocean inland varying distances that depend on topography. The moderating influence of the sea air is blocked by mountains, and gradually diluted by inland air. Away from the immediate influence of the ocean, the temperatures may reach extremes and are often unpredictable from one day to the next. For example, although uncommon, frost can occur any day of the year in the inland mountain chaparral, and snow may be frequent at times in winter (pl. 13). Large daily fluctuations in temperatures are common as well and are superimposed on the seasonal patterns. In the chaparral of central California, for example, it is not uncommon for the springtime dawn temperature to be freezing, whereas the noontime temperature reaches over 100 degrees F. Winter temperatures rarely stay below freezing for more than a short time, however, particularly in southern California. This is the reason that oranges and other citrus crops can

Plate 13. Snow-covered manzanita chaparral with cypress trees in the background (Guatay Summit, San Diego County).

be grown in the primarily gentle climate of the valleys below chaparral-covered hillsides (pl. 14). Occasionally, when the temperatures suddenly plummet in agricultural areas, smudge pots, fans, or electric heaters must be used to prevent freezing and destruction of the crop. It is advisable to have a sweater or jacket with you at all seasons of the year in the chaparral, as a warm spring afternoon can quickly become a cold evening.

Microclimates

Microclimates are variations in climate that occur at a small spatial scale. They are created by localized factors such as slope steepness, presence of rocky areas or ravines, and differences in the exposure of slopes on the south and north sides of mountain ranges. These physical factors affect the quantity of

Plate 14. Chaparral-covered hillsides behind the citrus orchards of the Ojai Valley, Ventura County. The vertical white structures in the orchards are wind machines, used occasionally on cold winter nights for frost protection.

sunlight and heat a slope receives, and therefore how moisture is distributed and retained. North- and east-facing slopes, for example, are shaded from direct radiation during the hottest part of the day, as are deep ravines. These sheltered places are consequently cooler and moister than south- or west-facing slopes or broad, open mountainsides. Because of a relatively high moisture level in these protected locations, the chaparral shrubs there are often taller and more lush than on exposed slopes. These are favored summer resting and feeding places for deer. Daily temperature fluctuations also tend to be less extreme on these more sheltered and moist slopes because the water in the air serves to buffer changes in air temperature. The contrast between shaded and exposed slopes can be significant, as can be seen in pl. 15. On adjacent north- and south-facing exposures, there are differences in the height, cover, and species composition of the shrubs. Sim-

Plate 15. Differences in climate on north- and south-facing slopes favor different chaparral shrub species. Chamise (in flower) dominates the south-facing slope on the left, while ceanothus and scrub oak dominate the cooler and moister north-facing slope on the right.

ilarly, rock outcrops provide shaded areas and protected recesses in the midst of open chaparral. They also affect local temperature and moisture conditions such that ferns and other delicate plants can be found in ledges, crevices, and at the base of boulders. These microclimates provide rich habitats for species of rodents, reptiles, and invertebrates that otherwise would be absent from mature chaparral. In addition, the crevasses and recesses afford shelter to animals that otherwise might be killed by wildfires.

Finally, two different environments in the mature chaparral are created by the plants themselves. The first one is quite visible: the dense, leafy shrub canopy of uniform height and interlacing branches that may be continuous for hundreds of acres in all directions, creating an uninterrupted surface. The air above and around these leaves is dry and hot during the day, cold and somewhat moister during the night. While there is some exchange of air around the leaves, for the most part this top layer takes the brunt of daily and seasonal temperature fluctuations. Below the exposed canopy, on the other hand, is a second environment where the microclimate is cooler because the ground is perpetually shady, and where variations in temperature and humidity are much less extreme than at the canopy. This space beneath the shrub canopy is dimly lit and bare except for a sparse leaf litter (pls. 2, 16). The activity of most animals is concentrated in this understory landscape (see chapter 5). After fire, the ground surface is exposed, resulting in a harsher environment, totally different from its prefire state.

Convergence

The overall effect of climate can be seen in the phenomenon of convergence. Convergence is the similar appearance of unrelated organisms due to common environmental conditions. The climate is the most important factor in shaping organisms' physical appearance. This is because climate determines the major physical factors directly impinging on an organism, such as the temperature, moisture, and seasonal patterns. For the five regions of mediterranean climate (California, central Chile, western Australia, South Africa, and the Mediterranean Basin), convergence in overall appearance of the shrubs is obvious even to the casual observer. Compare the other four mediterranean climate regions in pls. 17–20 to the photographs of California chaparral scattered throughout the text. A blue-green blanket of evergreen shrubs covers the hills equally, whether in Chile or Mediterranean Europe. In all of the mediterranean climate regions of the world the dominant plant species in many areas are drought tolerant, evergreen, sclerophyllous (hard-leaved) shrubs and small trees. Many of these also have the ability to resprout after a fire from roots, stems, or burls. The individual shrub species look alike superficially from one region to the next, even though they

Plate 16. The top of the shrub is usually hot and sunny, while the understory remains cooler and shady beneath this manzanita.

are not genetically related. For a thorough and well-illustrated discussion of convergence in plants of mediterranean climate regions worldwide, see *Plant Life in the World's Mediterranean Climates*, by Peter R. Dallman, listed in the Supplemental Readings and References section.

Plate 17. Mature fynbos, a chaparral-type shrub vegetation, found in the Cape Province, Republic of South Africa.

Plate 18. Matorral in the Gredos Mountains of central Spain. Chaparral-like vegetation grows in many places around the Mediterranean Basin.

Plate 19. This shrub vegetation of Western Australia has a similar appearance to that of California and other mediterranean climate areas.

Plate 20. The matorral of central Chile looks similar to the chaparral of California.

Convergence also occurs among animals specialized to a particular climate type. For example, species of kangaroo rats (Heteromyidae) of the California chaparral (see chapter 5) have a morphological and behavioral equivalent in the jerboas (Dipodidae) that inhabit arid parts of Eurasia and Africa. Both groups of animals use the same hopping form of locomotion, live in burrows, and depend entirely on seeds for their food and water. Other animals depend on houses or burrows to protect them during the day and forage in the shrubbery at night in similar fashion in each climate region. While each organism is a unique product of its genetics and environment, it is remarkable that the same solutions to the same environmental problems arise independently over and over again. Convergence is a testament to the power of the environment in shaping the plants and animals around us.

The following story illustrates how one chaparral insect responds to the sudden availability of water in an unpredictable environment. A second story about tarantula hawks (*Pepsis* spp.) in chapter 5 shows another way a native chaparral insect obtains scarce water and food in chaparral. These special adaptations represent fine-tuned responses to the variable climate of chaparral.

Rain Beetles Mate Only When There Is Rain

A success story of survival in the water-limited climate of the California chaparral concerns the rain beetles, a unique kind of scarab (*Pleocoma* spp.) well adapted to the extended droughts. Around 30 species of rain beetles are found in chaparral and woodlands between southern Washington State and Baja California. Adult females are olive shaped, about two inches long, and males are somewhat smaller and more streamlined. The oval body is shiny on top with dense hairs pointing downward on the underside (pl. 5). Only the males have fully developed wings. Rain beetles spend the greatest part of their lives as larvae in feeding tunnels deep underground, where they are protected from the rigors of heat and drought. The only time these insects are above ground is during winter when it is raining or snowing. Then the air is cool and moist enough that the rain beetles can maintain their water balance and avoid overheating. These beetles are specifically tuned to and dependent upon rain to complete their life cycles. The adult males emerge above ground only in response to rain, and then they fly as the raindrops are falling around them, even in a cold winter rain. Flying in the cold requires a great deal of energy because the males must maintain a body temperature almost as high as ours in order to fly well. Since adults cannot eat (they have no mouths) and must live on fat stored when they were larvae, a male on the prowl has only enough energy reserves to fly for a total of about two hours. Most flights occur at dawn and last only 20 minutes, so males potentially have about six days of rainy flying time in which to find a mate. Whether or not mating has occurred, once the male's energy supply is used up, he dies. When it begins to rain, males emerge at the surface in a matter of minutes. In search of a mate they fly back and forth across the chaparral close to the ground where the females wait at their burrow openings. Females signal their location to the males by emitting an airborne hormone, called a pheromone, as an attractant. When the male discovers the burrow of a female, he enters to mate. Once mated, the female plugs the opening to the burrow and returns to the depths of the soil. Here she waits until early spring when she lays eggs in tightly packed, fine soil. The eggs are large, up to .2 inches in diameter, and require a great deal of energy to produce. When her egg laying is finished the adult female dies.

The entire mission of adult rain beetles is reproduction. These insects do, however, have very long lives below ground. Rain beetle larvae are thought to live eight to 13 years in the soil, where they eat the roots of shrubs and trees. The larvae are also excellent diggers with strong jaws for chewing their way through soil. The larvae grow slowly, molting once a year until they reach maturity. In the final molt they become pupae that persist for a few weeks and then transform into the adult stage. The adults then remain underground until the rains come.

The complex life cycle of rain beetles is adapted well to the environment of the chaparral. Not only can rain beetles survive for long periods without the benefit of rain, but also they respond immediately when it does rain. Rain beetles have one further mechanism that lets them exploit the cold, wet times of the year even when temperatures are below freezing. Male rain beetles can vibrate the muscles of their thorax (the middle body section), thus raising their body temperature to the level necessary for flying. The thick covering of hairs on their undersides acts to preserve this heat in the same way that hairs help insulate mammals from cold temperatures. The hairs also keep wet, cold mud from sticking to their bodies. This kind of a load could weigh them down and require more energy for movement. The rain beetles are truly well adapted to rigors of life in the difficult climate of chaparral, reproducing when there is water and able to wait out years when there is not enough.

FIRE IS KEY to understanding chaparral communities. While most of the vegetation types of California are prone to fire, chaparral is particularly so. It is the large chaparral wildfires near urban areas that capture public attention, as they are experienced directly by those who live near the chaparral and indirectly by others who watch footage on television or read the news.

Though seemingly cruel and destructive from our perspective, fire is a natural and an essential part of the life cycle of the chaparral community. Fire recycles and rejuvenates, and without fire many of the commonly observed chaparral plants and animals would die out. There is a tremendous diversity of plants and animals in the chaparral, including many endemic species. These organisms depend on the heterogeneous environment and the shifting mosaic of habitats created by repeated fires (pl. 21). Chaparral vegetation and some of its botanical antecedents have existed for millions of years in California, and fire has always been an integral part of this community. In short, where there is chaparral there is fire.

Plate 21. Wildfires driven by erratic winds create a mosaic of burned and unburned chaparral, as shown in this photo from the city of Santa Barbara.

The Fire Cycle

The repeating pattern of fire followed by renewal and recovery of the chaparral community is called the fire cycle. This cycle and the resiliency of chaparral to fire are most obvious in the plants. Plant populations grow back and persist in the same place through repeated turns of the fire cycle, and they dominate the physical structure of the chaparral ecosystem. Animals track the plants' reestablishment and growth, waxing and waning in abundance with time in accordance with the changing structure of the vegetation (fig. 6).

Recovery of the chaparral shrubs after fire takes five to 10 years in most areas of the state. Initially, fire leaves behind exposed hillsides with blackened skeletons of shrubs protruding here and there, and with ash and charcoal on the surface of the soil (pl. 22). Most chaparral wildfires occur at the driest time of the year, late summer and early fall. Despite the dry conditions, recovery of the woody vegetation often begins immediately. Within just a few weeks after fire, green shoots can be seen at the base of burned trunks and stems (pl. 23).

The shrubs resprout after fire from reserves held in the root systems and burls. Burls, enlarged areas at the base of the stems, contain water and stored energy to produce new shoots and maintain the plant during the early phases of recovery. As a consequence, resprouting shrubs grow quickly, forming distinct clumps of stems one to three feet tall within a year (pls. 24, 25). All species of chaparral shrubs, with the notable exception of some species of ceanothus (*Ceanothus* spp.) and manzanitas (*Arctostaphylos* spp.), have a burl. Some burls, such as those of chamise *(Adenostoma fasciculatum),* are almost spherical, while others, such as those of silk tassels (*Garrya* spp.), are more like woody platforms. Even though burls lack a common anatomy, they all contain tissues capable of generating new shoots. Shoots arise from folds and recesses of the burl that are shallowly buried beneath the soil and pro-

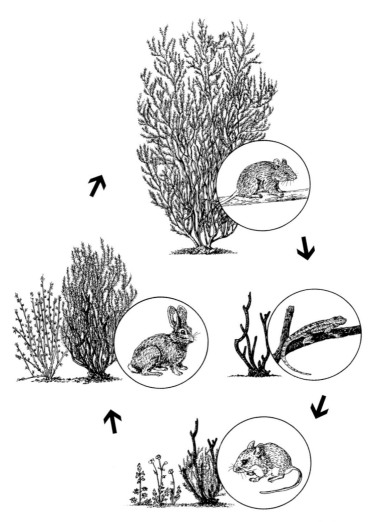

Figure 6. The fire cycle of the California chaparral. Fire destroys the dense shrub cover leaving hillsides blackened, but within a few weeks shrubs begin to resprout from burls or roots. The blackened branches make good perches for lizards to sun themselves and take advantage of the many insects that appear soon after fire *(right)*. Spring produces an abundant growth of flowering plants and resprouts, providing food and habitat for animals such as this Brush Mouse *(bottom)*. For the

Plate 22. A hillside blackened by a recent chaparral fire in the San Gabriel Mountains, Los Angeles County. The light patches on the dead branches of laurel sumac are caused by dead, peeling bark.

tected from the fire's heat. As shrubs grow older and survive repeated fires, the burls become progressively larger, and resprouting takes the form of a ring of shoots around the outer edge of the burl. Old burls can be as much as six feet across and are undoubtedly several centuries old. Some chaparral shrubs, such as laurel sumac *(Malosma laurina),* resprout not only from the burl, but also directly from roots (pl. 26). Not all individual shrubs capable of resprouting survive every chaparral fire, however. The proportion that does survive depends on the species, intensity of the fire, season of the burn, and health of the plants at the time of the fire.

In the first several months following a fire, the recovery of the chaparral vegetation accelerates. Winter rains dampen the soil, and the seeds of the fire annuals germinate, roused by the

next several years, subshrubs and recovering shrubs dominate, and Brush Rabbits and other small mammals become common *(left).* The shrubs grow to full size within a about a decade, returning the chaparral to its prefire appearance and providing habitat for wood rats and other animals of dense chaparral *(top).* Chamise is illustrated here at various stages of recovery from immediately after fire *(right)* to a mature shrub *(top).*

Plate 23. One season's growth of chamise sprouts from the basal burl, following fire.

fire from years of dormancy. These short-lived plants sprout in winter, gradually transforming the stark black hillsides to a soft green (pl. 27). As spring progresses, these annual plants often form dense stands producing spectacular displays of white, blue, yellow, purple, red, and orange flowers. The first spring after a chaparral fire is the most obvious manifestation of the regenerative power of the fire cycle and one well worth seeing (pl. 28).

After the initial display of the first spring after fire, the annuals begin to diminish in numbers and are replaced by low growing plants with woody bases and soft stems that are knee to waist high. These are called subshrubs. Subshrubs remain vigorous for roughly four or five years after fire, until the shrubs overtop them. Some of these plants are also brightly flowered and add to the beauty of the chaparral even after the annuals have faded.

Peeking out among the annuals and subshrubs, resprouting chaparral shrubs such as chamise, toyon *(Heteromeles ar-*

Plate 24. Resprouting shrubs such as this toyon, or California holly, can reach a large size within two to three years after fire.

Plate 25. A two-year-old burn site showing resprouting chamise along with interspersed subshrubs and sparse annual flowering plants.

Plate 26. Laurel sumac resprouts vigorously from its roots within a few weeks after fire in fall. Note that the surrounding ground remains bare, because of insufficient rainfall to stimulate germination of seeds.

Plate 27. Herbaceous plants grow quickly after a chaparral fire. The large flowering plant is Fremont's star lily, and the vine sprawling on the ground is wild cucumber. At the top of the photo, a coast live oak tree has new sprouts emerging from burned branches.

Plate 28. The first spring following a fall fire is one of the most spectacular sights in the chaparral, as California poppies and other colorful fire annuals flourish.

butifolia), and scrub oaks (*Quercus* spp.) appear as bright green tufts the first year. They darken with time, forming multistemmed clumps that steadily increase in size. Also, initially hidden by the annuals and subshrubs are the newly germinated seedlings of the shrubs. These shrub seedlings may remain concealed beneath the taller subshrubs and resprouts until they are several years old. Since reliably obtaining water is essential for a young shrub to become established, much of the initial growth goes into establishing the root system. Shrub seedlings grow down faster than they grow up. Most mature chaparral shrubs have deep and extensive root systems, so that what is seen above ground is a relatively small part of the plant (fig. 1, pl. 29). These deep roots hold the soil in place and on slopes are important in preventing erosion and landslides.

Growth slows when the shrubs become large enough to entirely cover the ground beneath, shading and crowding out

Plate 29. Chaparral shrubs are deeply rooted and hold the soil in place. The root of the laurel sumac to the right of the black and white measuring stick is 18 feet long from the base of the shrub to the point at which it disappears into the excavation hole. The complete root is considerably longer.

the subshrubs and herbaceous plants that may have persisted. After approximately five to seven years the chaparral is once again dense shrub vegetation. Then for long and indeterminate periods the shrubs continue to gradually increase in size, storing reserves in their burls and roots and producing seeds that then lay dormant in the soil until the cycle turns over again with the next fire. Since few chaparral shrub seeds germinate between fires, open patches may develop where large shrubs have died (pl. 30). This phenomenon is most common in chaparral dominated by ceanothus and along the lower margins of chaparral where it borders other types of shrublands or grasslands.

Animals also respond to the cycle of periodic destruction and renewal by fire. The time elapsed since fire is closely correlated with the composition of the species of animals that inhabit a particular stand of chaparral, because the age of a shrub stand governs key elements of habitat characteristics.

Plate 30. Shrubs of old chaparral stands may appear more variable and open than in younger stands as some individuals reach great size while others die back.

The quantity and quality of plant cover, the quality and variety of plant food, and even the potential locations and building materials for nests are all related to the time elapsed since the last fire. The open ground, newly available after fire, provides an opportunity for animals to exploit foods that were not destroyed by heat and flame. For example, kangaroo rats (*Dipodomys* spp.) depend on seeds buried in the soil before the fire. Other species, such as the harvester ants (*Pogonomyrmex* spp.), search for seeds blown in from beyond the perimeter of the burn. The Western Fence Lizard (*Sceloporus occidentalis*) uses newly available resources as it snatches up passing insects from its sunny resting place between burned stems of chamise (pl. 31). Mule Deer and Black-tailed Deer (*Odocoileus hemionus*) are also attracted to recently burned areas of chaparral, particularly when fires have been patchy, since this leaves open areas with lush vegetative growth adjacent to dense shrubs. Deer feed on the new growth of both

Plate 31. A Western Fence Lizard sunning on a rock among the burned stems of chamise.

herbaceous plants and chaparral shrubs while using the nearby dense vegetation for cover. Deer reproduction and survival are relatively high when burned and unburned areas are juxtaposed, and deer populations can increase tenfold in a few years time. As the recovering plant community generates a diverse and abundant mixture of succulent leaves, flowers, fruits, and seeds, the variety and density of insects, birds, mammals, and reptiles in chaparral increases very rapidly, reaching a peak two to five years after the fire. After two or three decades of postfire growth, the chaparral plant community becomes a structurally homogeneous habitat that supports fewer species of animals and excludes some that were present when the community was younger.

The Fire Regime

The overall pattern of recurrent fires in a particular place is called the fire regime. Four main elements contribute to the fire regime: frequency, intensity, seasonality, and spatial pattern. These elements are highly variable in many respects, contributing to the heterogeneity of chaparral in time and space.

Fire Frequency

Although the true natural frequency of fire is difficult to know, it is clear that fires are a periodic and inevitable feature of chaparral. The data on natural fire frequency are gathered from a number of different sources and they do not always agree. Fires have been observed to burn every five to 40 years in some areas, whereas they may occur only once a century in others. Chaparral grows in many settings, so that fire frequencies will always vary from place to place. Some factors that are important in natural fire frequency are species of shrubs present, elevation, latitude, proximity to other types of vegetation, frequency of ignition, wind patterns, topography, and time since last fire. Modern fire frequencies in many areas are affected by human actions (see chapter 6).

Fire-free periods of a century or more in chaparral watersheds are suggested by information obtained from charcoal varves in the Santa Barbara Channel. These varves are layers of sediment deposited annually on the sea floor from soil carried in the seasonal runoff of streams and rivers from the nearby Santa Ynez mountains. These layers contain charcoal if there has been a recent fire. Occasional years with large floods produce thicker varves than years with normal rainfall and runoff. Thick varves are also associated with floods that come from hillsides that have been denuded of chaparral by a recent fire, and these layers also contain charcoal. Consequently, vertical cores taken through the annual deposits can reveal patterns of fires and floods that go back for several centuries. The Santa Barbara Channel varves showed that in the Santa Ynez mountains there were two very large fire-flood episodes in the period between the years 1400 and 1550, and large fires somewhere in the area about once every 65 years over the past 600 years. Between episodes of large fires were quiet periods with few or no detectable fires. Similarly, records from soil profiles and tree-ring scars from Santa Cruz County in northern California indicate that large fires occurred there regularly prior to the European colonial period, at intervals from one to several hundred years.

Studies of fire frequency in San Diego County looked at the reproductive success of chaparral shrubs and trees with different intervals between fires. It was shown that short fire intervals could cause local extinction of chaparral plants. For example, a pair of fires a year apart wiped out the local chamise population. The first fire killed the tops of these shrubs, after which the population followed the usual pattern of quickly beginning to regenerate by resprouting and germination of seeds. A second fire the following year killed nearly all the seedlings and most of the resprouts along with their root systems. In another case, a fire interval of 30 years was also shown to be too short for the long-term survival of the Tecate cypress *(Cupressus forbesii)*, a small tree that grows among chaparral shrubs near the border with Mexico. Tecate cypress has cones that persist on the tree for years, remaining closed to protect the seeds within (a condition referred to as serotiny). Wildfires kill the mature trees but also cause the cones to open up within days, dropping the seeds onto the soil, where they germinate. Trees must be more than 30 years old to accumulate sufficient seeds stored within cones to produce seedlings at densities necessary for stand replacement after a fire. Similar work on seedling production by big-berry and Eastwood manzanitas (*Arctostaphylos glauca* and *A. glandulosa*) showed that more seeds were produced on 90-year-old plants than on nearby 25-year-old plants. The conclusion from these studies of shrubs and trees was that a fire-return interval of many decades, or even a century, is the optimum cycle for these plants. Long fire-free intervals pose no particular risk to chaparral shrubs and trees, but much shorter fire intervals could be very damaging. The death of some shrubs after several decades does not necessarily indicate that those particular species need fires at closer intervals. For example, populations of the shrub *Ceanothus tomentosus* that had largely died after 80 years in a chaparral stand were found to have left a sufficient store of viable seeds in the soil to produce a new population of seedlings after fire.

Studies suggesting more frequent chaparral fires in historical times come from newspaper accounts during early settlement days in Los Angeles, and more recent observations comparing fire frequencies between San Diego County and adjacent Baja California. Both of these studies concluded that in earlier times small chaparral fires in southern California were frequent and patchy. This pattern appears to have resulted from fires that burned until they were naturally extinguished. The Baja California study suggests that in the absence of fire suppression, the frequency of smaller fires in San Diego County would be higher than it is today, while the frequency of large fires would be less than it is now. This conclusion is not universally accepted. Both of these studies are described in greater detail later in this chapter.

Frequent chaparral wildfires are the rule in many areas now, although not from natural causes. People living and working close to chaparral are the ignition source of the majority of today's fires. According to U.S. Forest Service statistics, the most common cause of wildfires is children playing with matches. Other common causes are sparks from motorcycles without spark arresters, sparks from campfires, discarded cigarettes, downed power lines, accidents with vehicles and motorized equipment, structural fires, and careless burning. Events as seemingly unlikely as plane crashes, locomotive sparks, and birds electrocuted on power lines have started disastrous chaparral wildfires. These accidental sources are sometimes compounded by deliberate ignition, in other words, arson.

Fire Intensity

Fire intensity, another aspect of the fire regime, is also important in chaparral. A fire burning ferociously in a hot, dry windstorm will release much more heat in a minute or an hour than a fire burning the same vegetation under calm conditions and lower temperatures. Aluminum objects are frequently reduced to puddles in wildfires, and the melting point of aluminum is 1,150 degrees F. Under some extreme conditions even glass bottles are melted, requiring a temperature of 3,000 degrees F! These are peak temperatures, but even so, most chaparral wildfires have recorded temperatures at the surface from 350 to 800 degrees F. Although fire temperatures above the ground can be extreme, just a few inches below the soil surface it is much cooler. Seeds shallowly buried are protected from heat death by the insulating soil. Kangaroo rats and other animals with deep burrows are also very little affected by the fire burning above them (see chapter 5).

The intensity of the fire affects the microbiology of the soil, the amount of standing charred and uncharred wood, the amount of ash and remaining nutrients, and many other community characteristics. Very hot fires may reduce all the vegetation to ash, turning the soil surface to a loose powder devoid of organic matter and internal cohesion, so that it is directly exposed to the erosive forces of wind and rain. A cooler fire may leave many charred stems and living roots in place, slowing erosion and reducing the hillside slippage. Very hot chaparral fires turn much of the nitrogen from the plants and leaf litter into ammonia and other gases that escape to the atmosphere, so that less of this important nutrient remains in the soil to support the growth of new vegetation. Within the boundaries of a given fire, intensity varies greatly from point to point, depending on the distribution and quantity of fuel, wind patterns, topography, and other physical variables. In general, north-facing slopes burn more slowly and less intensely than exposed south- and west-facing slopes. Uneven combustibility of the shrubs may also add to differences in intensity.

Seasonality

The season of a chaparral wildfire is important. As summer wears on, chaparral shrubs become progressively drier and more flammable. The water content of the shrubs' leaves is depleted, and the roots are no longer able to extract water from the dry soil. The fine, dead branches inside mature shrubs become crisp, dry kindling. All it takes is a small source of fire, such as a match or spark, to set off an explosive blaze. Lightning strikes, the natural source of ignition, are most common in late summer and fall, making fires from this source more likely at this time than at other times of the year. The likelihood of a fire becoming large is also increased by gusty seasonal winds, such as Santa Anas, that are common in fall. High fire danger can continue until winter rains begin days to months later (see chapter 2).

Fires that burn between September and November, the most typical time for catastrophic wildfires, occur when shrub growth is minimal, and when few animals are engaged in reproduction. In spring, on the other hand, shrubs are actively growing and have moved most of their reserves of energy and nutrients from root systems to stems, leaves, and flowers. Fires during spring sometimes kill a substantial fraction of the shrubs that might have readily survived a fire in fall. This is because fewer reserves remain in the burl and the root system that the plant can use to regroup and begin growing again, and because of the lethal effect of steam in growing tissues. Many animals are building nests and raising their offspring in springtime, as well, so that the eggs and young of vertebrates and invertebrates are likely to be at their most vulnerable at this time. The soil is also likely to contain moisture at or near the surface where a fire's heat can turn water into life-destroying steam that sterilizes the uppermost layer of soil. For all of these reasons a spring fire, started by humans, can be much more damaging to chaparral plants, animals, and soil than even very hot fires that occur in fall.

Spatial Pattern

Pattern, another aspect of the fire regime, is created by the confluence of frequency, intensity, and season, plus topographic differences, local microclimates, the individual characteristics of shrub species, and the weather at the time. Fire burns differently in different vegetation types, and the chaparral varies as to the dominant species of shrubs in different regions. Each species of shrub burns in its own way. Shrubs do not gradually ignite and burn smoothly, but rather the shrubs burn one by one as the halo of hot gases volatilized from heated resins and waxes in the leaves suddenly causes the entire shrub to explode into flames (pl. 32). Since each species of shrub has its own particular characteristics of burning, chances are good that some individual plants on a hillside will be only singed while others will be incinerated right down to the soil. As mentioned earlier, cooler and moister north-facing slopes often burn incompletely, whereas hot, dry south-facing slopes are more likely to burn entirely. Furthermore, topographic features such as cliffs, canyons, and rock outcroppings can redirect the path and change the intensity of a fire.

Plate 32. Chaparral shrubs burst into flame from explosively ignited gases.

Chaparral areas also frequently border ravines and stream areas, and this, too, may cause different patterns of burning. For example, a fire that burns everything in its path when moving through chaparral shrubs may become a much less intense ground fire as it travels beneath a stand of live oak trees and may altogether skip the moist canyon bottom. Working together, all of these variables often result in islands of unburned shrubs next to areas that are completely blackened. Looked at on a large scale, we see a mosaic of different ages and sizes of chaparral stands across the state. Today, the mosaic of recently burned and long-unburned chaparral in California is made up mostly of big pieces. Largely, this is due to the size of modern wildfires. For more information see Fire Patterns in the Twentieth Century later in this chapter.

Sources of Ignition

The natural cause of fire in chaparral is lightning. In California, lightning strikes are most common in the interior mountains and foothills above 5,000 feet, where there are open forests and low, patchy chaparral. Consequently, fires started by lightning usually begin away from the coast but may spread in that direction if pushed by winds or storms. The peak of lightning activity is mid-July through September. Many of these naturally ignited fires burn out in a short time, leaving small, irregular patches. However, sometimes a fire does not go out, but continues to burn for weeks, depending on the weather and the fuels. Smoldering logs and litter can be a latent source of fire. During the nights following wildfires, the woody bases of large, burning chaparral shrubs can be seen glowing like jack-o-lanterns set out on blackened hillsides. If the weather turns dry and windy, flames can spring up anew from embers, igniting what may become a large and uncontrollable wildfire. Since gusty winds are a common feature of late summer and fall in many parts of the state, it is mostly during this part of the year that fires spread away from their initial source area and turn into large wildfires. In August 1977 a low-pressure area that stretched the length of California generated lightning without rain statewide. The lightning strikes ignited nearly a thousand wildfires in one night. One of them became the Marble Cone fire, which burned for weeks through much of the very old chaparral of the Central Coast Ranges and grew to be the third largest wildfire in the history of California. Very similar circumstances occurred again in August 1999, when several very large wildfires ignited by lightning burned until October in central and northern California.

Under most circumstances today, lightning is a minor factor in predicting the probability of chaparral wildfires. The millions of people in California provide a steady supply of ignition sources in and around chaparral, so that people have become the leading cause of chaparral wildfires. These human-initiated events are unlike those sparked naturally by lightning. Human-caused fires most often begin at low elevations near urban areas and along highway corridors rather than up in the mountains, where lightning is most common. In addition, there is little seasonal pattern to fires caused by humans, which may burn through the same area of chaparral more frequently than at intervals of decades or centuries, as happens from fires caused by lightning.

Aboriginal Burning

As an evolutionary force, fire has shaped chaparral over millions of years, and for most this time period humans were absent. For example, charcoal fragments of chaparral plants have been discovered in sedimentary rocks of the Santa Monica Mountains that are 20 million years old — long before humans arrived in North America. Across this expanse of time, populations of organisms that were unable to survive periodic wildfires were eliminated from ancestral chaparral, while those that possessed or evolved adaptations to survive wildfires persisted. The arrival of humans approximately 11,000 years ago brought about important changes in the fire regime by providing a new ignition source that could be applied any time the vegetation was flammable. Humans everywhere use fire in ways that modify the landscape, and the vegetation of fire-prone California was particularly susceptible to this kind of manipulation. It is hard to escape the conclusion that burning by humans has exerted a profound effect on the nature of chaparral and all the other fire-prone plant communities of California.

There is no doubt that the original inhabitants of California caused fires in many plant communities, including chaparral. Anthropologists and historians have made inferences about the influence of California Indians on the fire regime by examining the records made around the time of first European settlement in the second half of the eighteenth century. There is evidence of intentional burning of vegetation by Indians in both northern and southern California. A diary from 1774 by Fernando Rivera y Moncada, the Spanish commander of Alta California, noted regular burning to increase grass seed yield, and a 1793 account by naturalist José Longinos Martinez observed that Indians throughout Alta California were in the habit of burning brush to drive out rabbits, and to produce the edible vegetation that followed. The first law regulating fire in California was a proclamation in 1792 by Spanish Governor José Joaquin de Arrillaga, which prohibited deliberate burning of vegetation. He declared that he was acting against "the widespread damage which results to the public from the burning of the fields, customary up to now among both Christian and Gentile Indians in this country (Alta California)" (Clar 1959, 8–10). Some wildfires from this early period evidently were quite large.

Nineteenth-Century Fire

Written records from the nineteenth century also indicate the continued importance of fire as part of both the pristine and the inhabited portions of the state. For example, after a visit to Santa Barbara in 1831, Alfred Robinson wrote the following:

> A great fire had originated in the mountains to the south, which spread to the environs of the pueblo, endangering the fields of grain and gardens. It approached the low hills ... close to town and favored by a strong wind kept traveling along the mountain range. The sight was magnificent but terrible at night when the fire reached the rear of the town. Large cinders fell in every direction, even the very vessels in the harbor having their decks covered with burning ashes. The air was too hot to breathe. People fled from their houses to the beach.... When the fire came to the vicinity of the mission vineyard, its path was checked because of the green state of the vegetation but it continued its course to the mountains northward where everything was destroyed and for months afterwards the bare and blackened hills marked the course of the raging fire. (Robinson 1925, 130–131)

In 1836 Richard Henry Dana commented on the same fire, writing in *Two Years Before the Mast:*

> The town is certainly finely situated, with a bay in front and an amphitheater of hills behind. The only thing which diminishes its beauty is that the hills have no large trees upon them, they having been all burned by a great fire which swept them off about a dozen years before, and they had not yet grown up again. The fire was described to me by an inhabitant as having been a very terrible and magnificent sight. The air of the whole valley was so heated that the people were obliged to leave the town and take up their quarters for several days on the beach. (Dana 1949, 55)

The New Englander Dana made the same error as did many others who came to California later on. He assumed that fire was an unnatural event, and that were it not for fire, the surrounding hillsides would carry a forest resembling that which he had grown up with in Massachusetts. Dana undoubtedly saw the chaparral shrubs that would have covered the hillsides 12 years after the fire, but these stirred no further comment. He was, in fact, within sight of chaparral most of the time as he sailed up and down the California coast, although it is never mentioned in his famous journal.

Studies of the fire regime of the late nineteenth century on the chaparral-covered slopes of the San Gabriel Mountains, facing the Los Angeles Basin, show a different pattern of burning than we see today. An analysis of early newspaper accounts and descriptions from the first stewards of the newly created Federal Forest Reserves paint a picture of a landscape that was broken up by many small fires into patches of chaparral of different age classes. Fires started by summer lightning burned for weeks, or even months, sometimes until they were extinguished by rain in late fall. Most of the time these fires were scarcely noticed in the valleys because they were kept alive only by smoldering in the trunks and stumps of burned trees or by slowly creeping along the ground. Occasionally, these fires would flare into intense blazes like those that make dramatic television footage today. These bursts of flames occurred when smoldering fires were suddenly whipped into conspicuous blazes by the first Santa Ana winds of the season. A large fire that burns over many days and weeks, as these did, is actually many fires that are connected together in a time sequence by a common source of ignition. Such fires burned through many plant communities and many age classes of vegetation. As discussed in the following section, our current pattern of occasional large wildfires in chaparral and many smaller fires that are quickly suppressed may or may not be like that of the past, with or without aboriginal burning. The current regime is influenced by frequent ignitions and successful suppression of almost all fires shortly after they begin. Neither circumstance existed before the twentieth century.

Fire Patterns in the Twentieth Century

A study by Richard Minnich, comparing contemporary fire patterns in the chaparral of Baja California Norte, Republic of Mexico, with patterns in the same sort of chaparral just north of the international border in San Diego County, gives us one view of what the natural fire pattern might be even now in extreme southern California were it not for fire suppression policies (see chapter 6 for more explanation). The mean interval between fires was the same in both countries—approximately 70 years—but the burning patterns were quite different. In Mexico, most fires were relatively small and burned at moderate intensities, mostly in mature chaparral. These fires seldom burned very far, because they almost invariably stopped when they encountered nearby patches of chaparral that were younger and consequently less flammable. In contrast, during the same period in next-door southern California, there were far fewer fires, but a few of these were very large and burned with great intensity. This striking contrast between the behaviors of fires on opposite sides of the border was attributed to differences in fire suppression practices between the two countries. In Mexico there is little or no attempt to put out most chaparral wildfires, and they therefore burn until they go out naturally. These fires are usually small because the patches of older, highly flammable chaparral are relatively small and fire simply stops when it burns up against younger patches of chaparral. The patches are small because the numerous fires divide the chaparral up into small pieces.

North of the border in the United States, most fires are quickly extinguished, and large areas remain free of fire for several decades. This long period without fire allows extensive, contiguous areas to accumulate so much fuel that once ignited the fires are difficult to stop. Thus the fire regime in the chaparral of southern California appears to have been changed by fire suppression and with the resulting pattern of the chaparral across large areas. The mosaic of chaparral age classes created by fires in the United States is made of a few large pieces, while the mosaic in Mexico has a much finer texture of relatively small, numerous pieces.

An analysis of 90 years of fire records from the nine California counties between Monterey and San Diego with extensive shrublands by Jon Keeley and coworkers paints a different picture of the historical fire regime. Their study of fire records concludes that overall fire frequency increased during the twentieth century, as did the total area of chaparral and related vegetation burned in Riverside, Orange, and San Diego Counties. Contrary to prevailing opinion, the size of fires did not increase in this area that contains most of California's chaparral. An accompanying analysis of large shrubland wildfires driven by Santa Ana winds across the Santa Monica Mountains of Los Angeles and Ventura Counties showed that these fires burned through chaparral of all ages, not just older vegetation. They concluded that in extreme fire weather conditions, chaparral burns without regard to the age of the vegetation, so that managing fire hazards through chaparral fuel reduction is likely to be ineffective for the fires that burn during fierce winds, the very conditions that are most dangerous to life and property.

The management implication of Minnich's study is that suppression of all chaparral wildfires is self-defeating, because it ultimately results in very large and destructive conflagrations that cannot be controlled. The conclusion of Keeley and coworkers is that the very large chaparral wildfires that do the most damage have not increased in frequency. More fires are started and most are quickly extinguished, but the huge and unwieldy fires are natural and will continue to occur as long as chaparral is present.

Modern Fires

The ever growing danger of mixing chaparral with more and more human developments is evident from contemporary trends of wildfire size and destruction (pl. 33). The largest and most destructive fires in California are recent, and most of them were fueled by chaparral. Nine of the 10 largest wildfires in state history were in chaparral, and the other was in chaparral mixed with other vegetation. All but one of these huge fires occurred since 1970. They ranged in size from 117,000 to 280,000 acres, areas four to 10 times the size of the city of San Francisco. The Marble Cone fire of 1977 burned over most of the chaparral in the Los Padres National Forest of Monterey County. The Matilija fire of 1932 burned over much of the chaparral of Ventura County, traveling from the core of then remote mountains all the way to the sea.

In the fall of 2003 an arc of chaparral wildfires surrounded the vast metropolitan area of southern California. From the Mexico border to Ventura County, a complex of 10 fires blew against, around, and through the ring of spreading urbanization that has brought ever-larger numbers of people into a fire belt fueled by chaparral. In one week the flames consumed just under 700,000 acres, destroyed more than 3,700 homes, took 20 lives, and occupied over 14,000 firefighters. The combined property and other economic losses of those fires were estimated to be 3.5 billion dollars. The 281,000-acre Cedar fire burned across chaparral of San Diego County from the city of San Diego to the eastern backcountry, taking 14 lives and destroying over 2,200 homes, becoming the largest wildfire in the recorded history of California. Another very large chunk of San Diego County chaparral was burned by the 175,000-acre Laguna fire of 1970, which took 382 structures and 5 lives. Virtually all of the 2003 fires began in chaparral and were propelled by erratic Santa Ana winds. Some of these fires later spread into mountain forests. An estimated million tons of pollutants blanketed the region beneath a yellowish gray pall of smoke and rain of ash that kept children indoors and adults wearing face masks for a week, and eventually reduced visibility as far east as Texas. At the height of the fires a NASA satellite measured elevated levels of carbon monoxide a thousand miles out across the Pacific Ocean (pl. 12).

Exactly 10 years earlier a complex of 18 fires across southern California burned 189,000 acres, damaged or destroyed over 1,000 structures, caused 3 deaths and injured hundreds. A total of 30,000 people were forced to temporarily leave their homes. Damage from these regional catastrophes ran to billions of dollars—natural disasters on the same scale as those caused by severe earthquakes and hurricanes. These are examples of the worst of fire seasons. Over a 10-year period between 1985 and 1994 an average of 703 California homes were

Plate 33. Today most fires in the chaparral have a human rather than a natural cause.

destroyed by wildfires each year. Not all of these were chaparral wildfires, but most were. Given present trends of new settlement adjacent to and within flammable vegetation, the annual losses from chaparral wildfires are likely to increase. Most of these losses are not inevitable. The many measures that can be taken by individuals and communities to protect themselves against chaparral wildfires are discussed in chapter 6.

Natural Responses of Plants and Animals to Chaparral Fires

Notwithstanding the effects of different fire regimes, plants and animals in the chaparral possess adaptations that enable them to survive wildfires. Below are some examples of plants and animals with special adaptations to fire.

Fire Annuals

The seeds of many species of chaparral shrubs and herbs require fire to break dormancy and allow germination. This is very different from the seeds of plants in most other places in the world, which require water, light, and reasonable temperatures but are otherwise free to germinate at any time.

The chaparral wildflowers that appear in luxuriant displays after fire are largely restricted to the first spring after fire (pl. 28). Of the more than 200 species of short-lived herbaceous plants that grow after fire, only a handful are found at any other time. This means that many of chaparral's most spectacular species will be seen infrequently and only after fire (see also chapter 4). The seeds of these plants may lie in the soil for a century or more, invisible to us, if there has been no fire. These particular plants are referred to as fire annuals, or pyrophyte endemics. These unique chaparral plants absolutely require the burning away of the mature shrubs before they will germinate.

Seeds of fire annuals have special systems to detect the passage of fire. Some species respond only to the chemicals produced when the wood of the shrubs is charred, or to the gases given off from combustion. For some species the cue is so specific that once in contact with the seeds, germination takes place within 24 hours. The seeds are not fooled by an exceptionally hot summer, manual clearing, or human manipulations. These seeds can survive for very long periods of time. This is due in part to their low moisture content, which is less than a dry paper towel. Similarly, the seeds of most species of shrubs also require fire in order to germinate. For some species it is the cracking or burning away of the hard seed coat that is needed, but for many others it is a special cue that comes only from fire. The seeds of most species of shrubs also rest in the soil for many years, lying dormant until the next fire. Shrub seedlings are consequently rare in the mature chaparral but become abundant as soon as those shrubs are burned away.

Fire Beetles

As with the plants of the chaparral, fire is crucial for some species of insects to survive and reproduce. Some are also named for their close association with fire and are the animal equivalent of the fire annuals of the plant community.

Fire beetles (*Melanophila* spp.) are so named because they depend on burned and indeed burning chaparral to reproduce. They are specially equipped to detect a fire in progress and will fly from 20 miles away to meet each other at burning chaparral shrubs. It is only on these smoldering ruins that they mate and lay eggs. Fire beetles begin courtship while the bushes are still burning. The mating ritual allows time for the shrubs to cool off to the preferred egg-laying temperature range of 100 to 115 degrees F. These really are red-hot lovers.

They are known to seek out areas as hot as 800 degrees F on the way to their mating sites.

These fire beetles are flat-headed wood borers that prefer to live in damaged trees or shrubs, especially those killed by fire. They have unique sensors in their antennae that detect atmospheric heat and smoke at considerable distances, and heat receptors along the underside of the body. These receptors are associated with glands that secrete a special wax. This wax aids the beetles by coating their undersides as they become warm, preventing water from evaporating from their bodies. Interestingly, these wax glands are found only on the females and are thought to be correlated with the need for her to sit for a period of time on hot branches while laying eggs. Female fire beetles have long ovipositors (egg-laying organs) that pierce the wood so that eggs are placed in cozy places within the burned stems and branches.

Fire beetles are attracted to burning chaparral shrubs in enormous numbers. Thousands of iridescent black beetles arriving almost simultaneously can form a glistening film on the trunks and stems within minutes of the fire's passage. Mating and egg laying take only a short time, and the adult beetles may disappear again within a few hours, leaving no visible trace. The larvae that develop from the eggs eat the dead and damaged wood of the shrubs and may live in these woody stems for several years. As predictable and fascinating an event as the appearance of fire beetles might be, very little is known about their total numbers or locations during non-fire periods. Fires are sporadic and geographically unpredictable and so too are the populations of their fire-dependent insects.

Fire beetles are attracted not only to burning chaparral shrubs, but also to oaks and pines and to some rather unlikely targets that might also be in a fire area. For example, fire beetles may cause problems for fire fighters because the beetles land on these people instead of on the burned trees and branches. In an effort to attach to a slippery, sweaty neck or shoulder the beetles will use their jaws as well as their legs to hang on to the surface, and their bite, while not serious, can be annoying. The sensitivity of these beetles to heat and smoke is so acute that they can also be drawn away from natural areas to human gatherings such as football games, barbecues, and roof tarrings. Football games at the University of California at Berkeley in the 1940s and 1950s, when cigarette smoking was highly fashionable, were regularly affected by a rain of fire beetles dropping from the sky in search of a suitable place to lay eggs! At one time they were so common near hot and smoky cement plant stacks in Riverside and Ontario in southern California that they were referred to as "stack beetles." With improved air quality, these beetles are once again seen primarily when there are fires in chaparral and other natural vegetation.

Fire attracts other beetles as well, even though they do not have specific names that reflect this. For example, long-horned beetles (family Cerambycidae) are also attracted to burned and burning chaparral. Beetles of the genus *Tragidion* lay their eggs in the still-warm stems of burned chamise, scrub oak, and sugar bush *(Rhus ovata)*. Their larvae can be seen for several years after fire in these shrubs. *Xylotrechus,* another genus of long-horned beetle, uses holes in the bark created by fire as entry points to search for water and sugar in the still-living but now exposed shrub stems. They do not attack healthy shrubs but may serve to finish off those greatly weakened by fire.

SHRUBS MOST CLEARLY define the chaparral. These densely intertwined, multistemmed, woody, evergreen plants make up the continuous blue green blanket that appears from a distance to gently cover the hillsides. Up close, however, the shrubs grow in nearly impenetrable stands of tough drought- and fire-hardy plants. In ecological terms shrubs are dominant, shaping the physical and biological environment for all other species. Shrubs are also what burn so fiercely during a fire. The stems and leaves that accumulate over many years provide the fuel. Shrubs appear uniform over large areas because their life cycles are restarted with every fire, so all the plants in a given area are the same age.

Herbaceous plants appear in numbers only when the shrubs are burned away. Long unseen, fire-stimulated wildflowers and subshrubs emerge from seeds in the soil. In the first few years after fire, herbs (short-lived plants lacking permanent stems) and subshrubs (plants with soft upper stems that often die back to the woody, lower parts during summer) temporarily trade places with the woody shrubs as the most numerous and extensive plant types. This reversal of roles may last for the first three to five years after fire until the shrubs reassert their dominant place in the community. The shrubs maintain this dominance until the next fire (see The Fire Cycle in chapter 3).

Trees (plants with a single, thick, woody trunk and usually upward of 25 feet tall) are occasionally associated with chaparral, especially on moist slopes, in ravines, and at higher elevations. Some species of cypress and the bigcone Douglas fir are found embedded in the chaparral, but for the most part, the common oaks and pines of California do not form a significant part of the chaparral community.

Few chaparral plant species are found in other habitats or plant communities. They are different in this respect from many species of animals that use the chaparral for food, nesting, or other resources but also spend time elsewhere. One of the special features of the chaparral flora is habitat fidelity.

The plants of the chaparral are uniquely adapted to this habitat, whether their period of abundance is during the long fire-free periods when shrubs dominate, or confined to the period immediately after fire when the shrubs are no longer present. Seeds are eaten by many kinds of insects, birds, and mammals, and with few exceptions animals move seeds only short distances. Seeds that are not destroyed remain in the area to reproduce the same plant community again following the next fire. So, unlike the seeds of many plants, those of chaparral natives are not regularly dispersed far away from the parent plant.

In this book, we treat the common and dominant shrub species and shrub families of the mature chaparral first, as they are the most conspicuous plants and provide overall structure and organization to the community. Those trees that are likely to be encountered near chaparral are also briefly discussed. This is followed by descriptions of the common subshrubs and herbs and their families. These are the types of plants that dominate chaparral immediately after fire. California has an extraordinarily rich flora overall, and the chaparral is responsible for much of that richness. Drawings and photographs are provided to aid in recognizing the major chaparral plants discussed here.

An Evergreen, Shrubby Vegetation

The shrubby growth form and the dense and often impenetrable nature of chaparral are adaptations to the rigors of a mediterranean climate. These rigors include extended periods of heat and drought interspersed with rain of quite variable amount and duration. In addition, chaparral soils are often coarse textured and poor in nutrients. Water moves quickly through this soil so that the plants have only a short time to take advantage of the moisture. Most trees and broad-leaved plants need a regular source of moisture, moderate temperatures, and reasonably fertile soils to grow well. With the poor soils and unpredictable water supplies of the chaparral these criteria are not all met, and only specially adapted plants can make it their home. Fire adds another physical variable that also eliminates some types of plants that might otherwise overcome obstacles to life in the chaparral.

Chaparral shrubs are woody, tough, and once established, capable of hanging on in difficult environments. Shrubs do not require as much energy to live as do trees, because of their smaller stature and more slender stems and branches. In addition, should one of the main stems prove to be in a bad location relative to water or sunlight, it may be allowed to die (self-pruning) without killing the entire plant. The many dead stems found beneath and within the mature chaparral canopy attest to this harsh but effective accounting system. The living tissue in the main woody trunks and stems is often reduced to only small areas as a result of the successive loss of branches over time. These shrubs also quickly develop deep root systems and can persist and indeed flourish on exceptionally steep slopes (pls. 10, 29). In the chaparral many shrubs have an enlarged woody base, called a burl or root crown, from which the main stems emerge. As mentioned earlier, the function of the burl is to produce new shoots after damage by fire or other causes. It is one of the features that make the chaparral shrubs so resilient.

The leaves are the energy producing part of the shrubs. It is their job to perform photosynthesis and make energy available as sugars and other carbohydrates for growth and reproduction. To do this they must have sunlight, water, and carbon dioxide and be able to exchange gasses with the air. This means that leaves are also the places where the demand for water is greatest and where the most protection is needed to avoid drying out. Chaparral shrub leaves do not wilt and rarely show visible signs of water loss even though they can be very dry (less than 10 percent water). Conditions favorable for high photosynthesis rates are not present during the entire day or at all seasons. Herbivores nibbling the leaves, and water lost through the surface further decrease productivity. To protect themselves from the costs of repair associated with wilting or mechanical damage, chaparral plants have tough, rigid leaves that are structurally reinforced with hard, non-photosynthetic tissue (sclerenchyma). This heavy reinforcement makes hard (sclerophyllous) leaves (pl. 34) that retain their structural integrity through the seasons. The typical description of chaparral, in scientific publications, is an evergreen sclerophyllous shrub vegetation. "Evergreen" refers to the fact that the leaves persist for more than one growing season and typically remain green throughout their life on the plant. This is different from deciduous plants, which shed their leaves during fall or when environmental conditions are unfavorable and then produce new leaves the next growing season. Chaparral shrub leaves may live for one to seven years, depending on the species and the plant's microclimate.

Chaparral shrubs grow so close to one another that the branches of adjacent plants are interlaced. These densely in-

Plate 34. The leaves of chaparral shrubs, like this lemonadeberry, are evergreen and hard. They are often oriented vertically to reduce heating by the sun and may have waxy or hairy surfaces to retard evaporation. A cluster of flower buds sits among the stiff, thick leaves.

tertwined branches shade not only the ground but also the plants themselves. Twisted branches at many levels within the canopy, combined with varying angles of exposure to the sun for vertical or fasciculate (clustered) leaves, means that not every leaf will be in full sun all the time (pl. 41). The partial shading of leaves can be important in this very sunny climate. The leaves actually perform more efficiently during the summer months with less than maximum exposure to the sun.

All organisms live between their limits and within their tolerances. The plant growth form that meets these specifications in a mediterranean climate is the woody, evergreen shrub. This common solution to the rigors of the environment can be seen clearly in the five areas of mediterranean climate around the world. Whether in Western Australia, South Africa, Chile, California, or Europe, many parts of mediterranean climate regions are dominated by evergreen sclerophyllous shrubs (pls. 17–20). The water-loving plants we commonly use to landscape our gardens and public spaces cannot survive without frequent and regular watering in this climate.

Common Shrubs and Shrub Families

More than 100 species of evergreen shrubs occur in the chaparral statewide, but only a few are common throughout. Also, only a small number of plant families contain most of the chaparral shrubs. It is, therefore, relatively easy to recognize most of the widespread shrubs of the chaparral, such as chamise, manzanita, and ceanothus. This is not the case in many other places in the world, and it is a distinguishing characteristic of the California chaparral. The most abundant and commonly encountered chaparral shrubs are found in five families. These are the rose, the buckthorn, the heath, the oak, and the sumac families. Several other plant families occur in the chaparral, as well, and can be locally common. These less widespread families are discussed following the review of the five that are most commonly represented.

The Rose Family (Rosaceae)

The rose family contains several of the most common and widespread shrub species in the chaparral. As a group, species of chaparral roses are the tough, resilient, woody survivors of an ancient vegetation that covered much of the western United States millions of years ago. These include the plant that most characteristically defines the California chaparral, chamise, and the plant that gave Hollywood its name, California holly, or as it is more commonly called today, toyon. Other chaparral members of the family include the holly-leafed cherry, mountain mahogany, and a cousin of chamise, red shank. Outside of the chaparral this family includes economically important plants such as apples, cherries, peaches, plums, pears, strawberries, raspberries, and of course, roses.

Chamise

The most abundant and ubiquitous shrub of the chaparral is chamise *(Adenostoma fasciculatum)* (pl. 35). It has a range that encompasses almost the entire latitudinal spread of chaparral from northern California to Baja California. The cover photograph shows it growing near the northern limit of its distribution, mixed with other species of chaparral shrubs and trees. It is also the most common species of shrub in over 50 percent of the chaparral statewide. It often grows in almost pure stands where very few shrubs of other species are present. The average chamise shrub is four to six feet tall. It can be readily recognized by its small (one-half inch or less), dark green, needlelike leaves grouped together in numerous clusters (called fascicles) dotted along the pale gray stems (pl. 11, fig. 7). Usually, several stems arise from a burl at the base of the shrub. This species resprouts vigorously after fire, and an individual plant may survive for hundreds of years.

Plate 35. Chamise, a member of the rose family, is the most common shrub in the chaparral throughout California. Here, blooms are seen on near and far slopes.

Chamise has a short growing season, starting in late February or March and ending after the rains, usually sometime in June. Growth is from the tips of existing branches, and new lengths of growing stem can be seen readily in spring. The reddish tinge and soft texture of the new growth clearly distinguishes it from the harder, gray-barked branches of prior seasons. The plants flower in early summer, after their spurt of growth. The flowers are small with creamy white, frothy-looking blossoms densely packed into clusters up to eight inches long at the ends of the branches. A June hillside of flowering chamise transforms the appearance of the chaparral from green to white, because countless flowers temporarily obscure the foliage (pl. 35).

The creamy white of new flowers is replaced by a rusty red brown as the seeds develop and flower parts dry out, as seen in the foreground of the book's cover photograph. The shrubs are thickly covered by millions of flowers during spring, but only a small percentage of the seeds produced are capable of germinating. Of those that are viable, fewer still will ever have

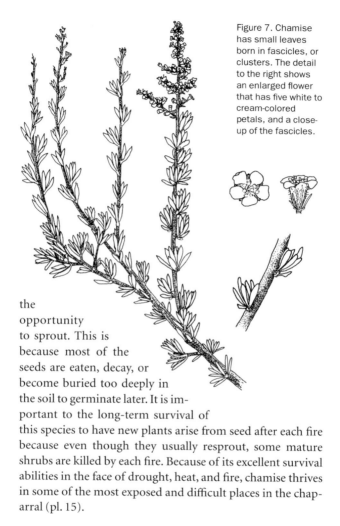

Figure 7. Chamise has small leaves born in fascicles, or clusters. The detail to the right shows an enlarged flower that has five white to cream-colored petals, and a close-up of the fascicles.

the opportunity to sprout. This is because most of the seeds are eaten, decay, or become buried too deeply in the soil to germinate later. It is important to the long-term survival of this species to have new plants arise from seed after each fire because even though they usually resprout, some mature shrubs are killed by each fire. Because of its excellent survival abilities in the face of drought, heat, and fire, chamise thrives in some of the most exposed and difficult places in the chaparral (pl. 15).

Red Shank

Another member of the genus *Adenostoma*, red shank *(A. sparsifolium)*, also called ribbonwood, is found over a narrower geographical range than chamise. It occurs only in central and southern California, from San Luis Obispo County in the Central Coast Ranges to the southern limit of chaparral in the Sierra San Pedro Martir of Baja California. Within these limits, it seems to replace chamise in places with more moisture or better soil. Red shank tends to grow in pure stands, surrounded by chamise, or by manzanita (*Arctostaphylos* spp.) and ceanothus (*Ceanothus* spp.) on the higher slopes. Curiously, it does not naturally occur in the Transverse Ranges of southern California, although when it is planted there among chamise and other species of chaparral shrubs it grows quite well and regenerates after fire. Red shank and ribbonwood are names given to this plant because of the loose shaggy red bark that peels off in long strips from the major stems. The shrubs are tall by chaparral standards, often growing to heights of 12 to 20 feet (pl. 36). The leaves are similar in size to those of chamise and are conspicuously resinous (fig. 8). They grow in linear clusters at the end of the branches, in loose broomlike

Plate 36. Red shank, shown here in flower, is a close relative of chamise. It is a remnant of an ancient vegetation that existed in southern California before the modern chaparral.

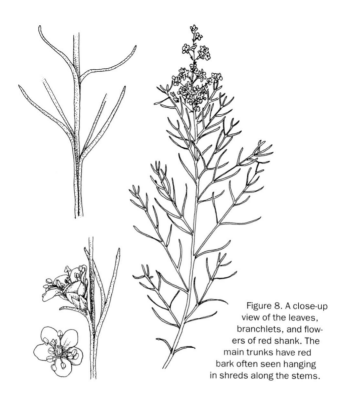

Figure 8. A close-up view of the leaves, branchlets, and flowers of red shank. The main trunks have red bark often seen hanging in shreds along the stems.

sprays. Given the height of these shrubs and the leaf arrangement at the ends of the branches, it is sometimes possible to walk upright beneath the long, wandlike branches. Red shank has many small, creamy white flowers, similar to those of chamise, but in open clusters six to eight inches long. These may be dense at the ends of individual branches but rarely hide the foliage as chamise blossoms do. As with chamise, a late spring flush of new leaves and stems is followed by flowering and seed production. Red shank resprouts well and also produces new seedlings after fire from seed stored in the soil.

Toyon

Another member of the rose family, toyon *(Heteromeles arbutifolia)*, also called Christmas berry or California holly, is one of the best known and most widespread of all chaparral shrubs. This is the plant that gave Hollywood its name, and it remains a common shrub in the Hollywood Hills today (pl. 4). Toyon is found in chaparral from northern California to Baja California and from the Coast Ranges to the foothills of the Sierra. This species favors cooler east- and north-facing slopes and canyons in the chaparral. Toyon grows as scattered individuals among other shrub species rather than in pure stands. This is likely due to birds eating the berries and later dropping the seeds randomly among other shrubs. Toyon is often one of the tallest chaparral shrubs, reaching a height of up to 20 feet. It has sclerophyllous evergreen leaves with serrated and bumpy margins, but the leaves are not prickly like those of holly (fig. 9). The mature leaves of this species are two to four inches long and about a third as wide and may live as long as seven years under good conditions. One characteristic of the young leaves is a reddish color and a characteristically bitter odor when crushed. This odor is due to the presence of cyanide-containing chemicals that are accumulated there by the plant to deter herbivores from eating the tender foliage. Despite this, tender young toyon leaves are a favored food of Mule Deer *(Odocoileus hemionus)*.

Flowering takes place in the late summer after a spring flush of new leaves has been produced. The small white flowers are produced in broad flat clusters up to 10 inches across and scattered over the plant. The characteristic pom-poms of red berries (pl. 4) develop over several months and are most conspicuous in winter. The common names California holly and Christmas berry were given to this shrub because of its superficial resemblance to English holly *(Ilex aquifolium)* with its winter clusters of bright red berries and contrasting dark green leaves. Early immigrants with English roots used it in

Figure 9. Foliage, flower, and fruit cluster of toyon, a common shrub in moist canyons throughout the state.

their California homes for "traditional" Christmas decorations. The practice of decorating with the berries and branches became so widespread and destructive in the Los Angeles area during the 1920s that a state law was passed prohibiting the harvesting of this plant. Toyon, its other common name, is derived from an old Spanish term for "canyon," the habitat preferred by this species. It is a vigorous resprouter after fire (pl. 24) and also produces seedlings. Unlike many other chaparral species, seeds of this species can and do germinate between fires.

The one-quarter inch diameter toyon berries have the

same structure as tiny apples, and they are edible by humans after they are heated to remove the bitter taste. Both Indians and early settlers cooked toyon berries lightly by boiling, steaming, or roasting. Mexican and American settlers made a cider from the berries, and Indians used an extract of the bark for various aches and pains. Early Spanish settlers made a sort of pudding by baking the berries slowly with sugar and used this as a pie filling. These fruits are highly prized by numerous species of birds, as well. They will often flock to the same bush day after day in winter until all the berries have been consumed.

Toyon was introduced to cultivation in Europe 200 years ago, shortly after Europeans began exploring the flora of California. It is also widely grown in parks and gardens throughout California today.

Mountain Mahogany

A prominent species of steep slopes and rocky, dry chaparral is mountain mahogany *(Cercocarpus betuloides)*. It is found from Oregon to Baja California, in the cismontane of southern California, and out to the Channel Islands. Individuals may range from six to 20 feet tall. Typically the shrubs are scattered, with visible space between individuals when they are found on steep slopes, but this species may grow in dense aggregations in other areas. Mountain mahogany is often found in association with the chaparral shrubs chamise, ceanothus, and manzanita. Mountain mahogany leaves, one-half to one inch long, are ridged and grooved in an almost fanlike pattern and are dark green on the top surface, with white, hairy lower leaf surfaces. The stems are also almost white and easily seen because of the sparse nature of the leaves. It has white flowers about one-quarter inch across, spaced along the upper branches (fig. 10). Flowering may start in early spring and continue through May, when the fruits and seeds mature. While flowering often makes chaparral shrubs conspicuous, it is the fruiting season of mountain mahogany that calls most attention to this plant. Thousands of pale feathery plumes extend out from the branches all over the outside of the shrub canopy and, when backlit, give the shrubs a silvery halo. The bottom of the spiral plume is attached to a seed, which is enclosed in a structure that resembles a pointed drill tip. When the fruits eventually drop to the soil they work their way into the ground as the wind pushes the spiral plume around in a circle. The seed is actually screwed into the ground. Once the seed is embedded in the soil, the feathery top breaks off and blows away. Mountain mahogany resprouts and also reproduces from seed after fire. Two other members of the genus are desert mountain mahogany *(C. ledifolius)*, found in particularly dry areas in the mountains and in desert chaparral, and small-flowered mountain mahogany *(C. minutiflorus)*, a species restricted to chaparral of San Diego County and Baja

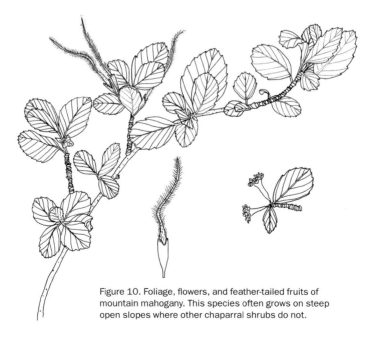

Figure 10. Foliage, flowers, and feather-tailed fruits of mountain mahogany. This species often grows on steep open slopes where other chaparral shrubs do not.

California. The foliage of mountain mahogany is a favored food of deer as well as domestic browsing animals.

Holly-leafed Cherry

In contrast to mountain mahogany, holly-leafed cherry (*Prunus* subsp. *ilicifolia*) prefers ravines, north-facing slopes, and relatively moist environments within the chaparral. This species is found in the chaparral of the Coast Ranges from Napa County south to Baja California and on the Channel Islands (fig. 11). Holly-leafed cherry is in the same genus as the cultivated cherry. It can range from a shoulder-height shrub to a treelike form 20 feet tall, usually with more than one major trunk. It has ruffled, shiny, dark green leaves three-quarters to two inches long with prominent spiny margins. The small white flowers are born in branched sprays (racemes) several inches long and can be seen in spring and early summer. When the sweet tasting fruits ripen to a dark purple they are clearly recognizable as cherries, although often smaller than the commercial variety. The fruits are much sought after by birds. Holly-leafed cherry resprouts vigorously after fire but like toyon does not require fire to reproduce. Seedlings are occasionally observed in openings and canyon bottoms in areas of mature chaparral. A subspecies of this species, the Catalina cherry (*P. ilicifolia* subsp. *lyonii*), from the Channel Islands, is notable for its height and bright, shiny green leaves. These features make this variety popular as an ornamental plant along the coast and in the valleys of California.

The Buckthorn Family (Rhamnaceae)

The buckthorn family is abundantly represented in the California chaparral, particularly by species of ceanothus, also called California lilac. The chaparral is the center of ceanothus worldwide. Many endemic species have evolved in particular chaparral locations and on special soil types. One of the most interesting questions to students of the chaparral is why there

Figure 11. Holly-leafed cherry has two geographical forms, or subspecies. The smooth, large-leafed form *(above)* comes from the Channel Islands and is commonly called Catalina cherry. The spinier and smaller-leafed form *(below)* is usually referred to simply as holly-leafed cherry and is found in the chaparral throughout mainland California.

are so many species of ceanothus in California and the chaparral when there are few elsewhere. Much remains to be discovered about these species, as the answer is not yet known. Nearly every part of the chaparral has one or more species of

Plate 37. Ceanothus covers broad expanses of chaparral in the Santa Ynez Mountains above Santa Barbara. The paler ridges and upper slopes show the cream colored flowers of bigpod ceanothus. Chamise mixed with manzanita covers the hill in the immediate foreground.

this genus, making ceanothus as characteristic of the chaparral as the ubiquitous chamise (pl. 37). Other members of the buckthorn family, spiny redberry, holly-leaf redberry, and California coffeeberry, are also found in the chaparral across the state, often in combination with ceanothus. This is a hardy family in the chaparral. All of the buckthorns can form dense, wiry, and often spiny thickets, making them among the most difficult parts of the chaparral to penetrate.

Ceanothus

Along with chamise, ceanothus (*Ceanothus* spp.) is found in nearly all chaparral areas throughout the state. Of the more than 40 species in the chaparral statewide, nearly every part of the chaparral will have at least one species of ceanothus and generally more. A given hillside may be dominated by one species or have a patchy mixture of several species often mixed with the ubiquitous chamise. Each ceanothus species prefers a slightly different location along the slopes. Species also grow at many different elevations and exposures across the state. Ceanothus species vary collectively in size from dense low-growing mats to individuals 15 to 30 feet tall. The leaves of ceanothus species can be from one-half to two inches long and vary from relatively smooth, uniformly pale-colored types to tough, thick ones that have a bumpy, wrinkled appearance and may be strongly bicolored dark green and white (figs. 12–14). The leaves are important in distinguishing the two main groups of species within the ceanothus genus. The characteristic that gives ceanothus its other common name, California lilac, is profuse clusters of sweet-smelling white, blue, or purple blossoms that cover the shrubs in late winter and spring (pl. 3). The individual flowers are quite small, about one-tenth inch across, but occur in large clusters that are superficially similar to lilacs when viewed from a distance. Each species of ceanothus produces flowers over approximately a period of two months between January and June. All the members of a particular population of ceanothus tend to flower at once, and each species blooms according to its own internal clock. This produces a patchwork landscape with splashes of color that come and go over a period of several months, creating a traveling kaleidoscope of creamy white and blue, sequenced by time, elevation, and species. These displays are sometimes bright enough to be seen from an airliner. The heavy, sweet odor of countless flowers can be detected for long distances, and ceanothus nectar is an important food for honeybees (*Apis* spp.) and many other insects.

Ceanothus fruits are explosive when ripe. The explosion occurs from tension that builds within the structure of the three-chambered fruit as it dries out in warm weather. There is enough force to make a clear popping sound and to launch seeds in all directions with considerable force and velocity. The simultaneous explosion of many capsules creates a small

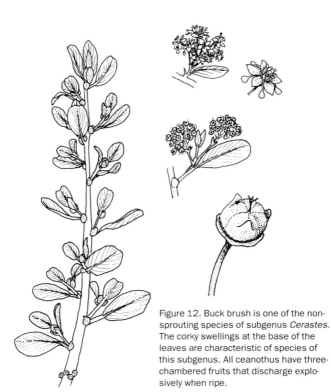

Figure 12. Buck brush is one of the non-sprouting species of subgenus *Cerastes*. The corky swellings at the base of the leaves are characteristic of species of this subgenus. All ceanothus have three-chambered fruits that discharge explosively when ripe.

storm of flying seeds. The newly released seeds of most ceanothus species are dark and shiny so that the ground looks conspicuously peppered with them. They are also numerous. For example, beneath hoaryleaf ceanothus *(C. crassifolius)* (fig. 14) in a good year there may be 10,000 to 12,000 seeds per square yard! Initial seed densities, even when quite high, drop steeply in a matter of few weeks. Those seeds that escape the keen eyes and noses of the industrious rodents, birds, and ants are buried in the soil, where they become part of the soil seed bank and await the next fire for an opportunity to germinate.

Figure 13. *Ceanothus tomentosus* has white to light purple flowers and shiny, dark green leaves. Characteristic of members of the resprouting subgenus *Ceanothus,* the leaves of this species have three main veins.

Ceanothus species are capable of adapting to a wide range of conditions. This ability is valued in the production of horticultural hybrids, meaning crosses between different species. About 20 species of ceanothus are popularly planted as ground cover, on steep slopes, for garden contrast and color, and along roadsides. Hybrids can be especially suited for a particular region's climate and soil, as well as valued for their decorative nature. As a group, ceanothus grows in almost any habitat in the mediterranean zone of California with full sun and well-drained soil, requiring little or no water beyond that provided by nature. In addition to the ability to hybridize

Figure 14. Foliage, flowers, and ripening fruit *(center right)* of hoaryleaf ceanothus, another nonsprouting member of the subgenus *Cerastes*. The leaves are small, hard, and densely hairy on the underside.

readily, ceanothus is also tolerant of a wide range of soil types, including serpentine soils, which are poisonous to most other plants. This tolerance is thought to be an important factor in the formation of so many species in different habitats throughout the state. Half of the 40 California species of ceanothus are narrow endemics, in extreme cases limited to a single site. Examples of very narrow endemics are the endangered Pine Hill ceanothus *(C. roderickii)* and Coyote Valley California lilac *(C. ferrisae),* limited to serpentine outcrops in El Dorado County and Santa Clara County, respectively.

Eighteen ceanothus species are on the Species of Special

Concern list compiled by the California Department of Fish and Game as part of the Natural Diversity Database (NDDB). This list contains species and subspecies that are rare, limited in distribution, or legally listed as rare, threatened, or endangered by the state of California or the federal government. Because of the narrow distribution of many of these species, habitat destruction is one of the major problems for surviving individuals.

Taxonomists divide the ceanothus genus into two groups: subgenus *Ceanothus* and subgenus *Cerastes.* These two groups differ from each other in two broad characteristics: the leaf type and the ability to resprout after fire.

The subgenus *Ceanothus* comprises species that have leaves with three prominent veins. The leaves are arranged alternately along the stem (fig. 13). These species also have the ability to resprout after fire. Leaves vary from a smooth, almost succulent condition, to hard and roughly textured. In extreme cases, the texture of the leaf may make it difficult to identify the three veins. In this case it is best seen by looking at the underside of the leaf, where the veins are more prominent. The stems are often a smooth pale gray or green near the leaves but become rougher and more conspicuously brown and gray as the bark thickens near the stem base. The burls are sometimes irregular in shape and may not be obvious on steep slopes, where they can be buried by loose soil and litter. Some examples of a resprouting species are chaparral white thorn *(C. leucodermis)* (pl. 38), common in southern California, coast white thorn *(C. incanus)* of the Central Coast Ranges, and tobacco brush *(C. velutinus)* of northern California.

The members of subgenus *Cerastes* have only one main vein running down the center of the leaf, and the leaves grow in pairs directly across from one another on opposite sides of the stems. The adult plants of these species are killed outright by fire, and they do not have a burl. The leaves are usually quite thick in these species and are often strongly bicolored,

Plate 38. Chaparral white thorn, a resprouting species of ceanothus, produces abundant clusters of flowers during the spring.

with dark green upper surfaces and pale to white, hairy lower surfaces that may also be dotted with small pits. These leaves may appear to roll up into cylinders and become almost vertical as the environment becomes drier through summer and fall. The stems are a pale gray and have pronounced darker corky bumps just below the leaves. These bumps, technically called stipules, persist on the older twigs and branches so that they may be seen even after the leaves fall off. The function of these structures is unknown (figs. 12, 14).

Another highly characteristic feature of nonsprouting species of ceanothus is the braided appearance of the main trunks and stems. This braiding is caused by the death of some parts of the trunks while others remain living and grow around the dead areas. The mix of living and dead areas produces a dimpled or ropy effect. This appearance, called longitudinal bark fissioning, is part of the overall survival strategy of ceanothus. If a root is not in a good location to obtain water, it will die. The part of the stem to which it is connected will also die. Similarly, if a branch is too deeply shaded or otherwise unable to do enough photosynthesis to "pay" for the cost of maintaining it, the branch will die and so too will the root attached to it. The chaparral is a difficult environment for these plants, especially during the dry season, and the only way to survive is to drop the parts that are not cost-effective. The characteristically braided appearance of the stems can still be recognized in the remains of burned stems after fire. While most chaparral shrub species live for great lengths of time, many ceanothus species, especially those that are non-sprouters, appear to have a more limited life span. Places that have not burned in 50 or more years may have more dead than living ceanothus shrubs. Factors contributing to longevity are not well understood. Some of the more commonly encountered nonsprouting members are bigpod ceanothus *(C. megacarpus)* (pl. 3) and hoaryleaf ceanothus *(C. crassifolius)* (fig. 14) in southern California, while buck brush *(C. cuneatus)* (fig. 12) and wavyleaf ceanothus *(C. foliosus)* are common in the central and northern parts of the state. Members of both subgenera produce seedlings after fire from seeds stored in the soil, but for nonsprouting species, this is the only way to survive. Consequently they are sometimes referred to as "obligate seeders."

Ceanothus species, both sprouting and nonsprouting, possess an important ability that allows them to grow on particularly poor soils. They are among a special category of plants called nitrogen-fixers. Commercial fertilizers include a healthy dose of nitrogen-containing compounds, and it is the nitrogen that gives fertilizer its strong smell. Inside the roots of ceanothus species, nitrogen from the air is changed chemically into a useful form by bacteria that live within special pockets in the roots, called root nodules. Since these bacteria are housed within the roots, the nitrogen can be pumped directly into the plant as it is produced. The relationship between the bacteria and the ceanothus is mutually beneficial to both species, as the bacteria receive a place to live while the plant receives nitrogen in a useful form. Measurable enrichment of the soil with nutrients such as nitrogen takes years in the stressful environment of the chaparral and is another area of chaparral ecology in need of more study.

Spiny Redberry and Holly-leaf Redberry

Two other buckthorn family members of the chaparral are spiny redberry *(Rhamnus crocea)* and holly-leaf redberry *(R. ilicifolia)*. Both redberries (fig. 15) are found from the North Coast Ranges to Baja California, usually below 5,000 feet elevation. These shrubs range in height from five to ten feet tall and have shiny, spine-tipped branches and twigs one-half to one inch long. Spiny redberry is most often found where it is sunny, dry, and hot for most of the year and tends to the lower end of leaf size for chaparral plants. Holly-leaf redberry flourishes in moist ravines and shaded locations. Shrubs may be dense with many lateral branches that sometimes retain foliage to the ground level. Flowering occurs in March and April. By the end of spring the shrubs are conspicuously dotted with small, shiny red berries. Both species resprout after fire. Seedlings are most common after fire, but plants may occasionally establish from seed between fires in cleared areas. These two species are very similar in appearance and are sometimes treated as varieties of a single species, rather than as separate species.

California Coffeeberry

A common chaparral shrub in the buckthorn family, California coffeeberry *(Rhamnus californica)* is named for its red to black fruits, which are the size of coffee beans (pl. 39). This species is widespread and is found from northern California to Baja California in the inland and coastal ranges. Individuals range in height from six to 25 feet tall, growing especially large in moist areas. The elliptical leaves vary from two to five inches in length. In spring and when there is abundant moisture, the leaves are light green, positioned horizontally along the branches, and almost fleshy in comparison with most

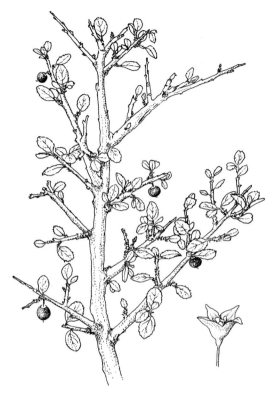

Figure 15. Spiny redberry, a member of the buckthorn family, derives its name from the many small bright red berries that dot the branches in late spring. The short thornlike twigs and the dense growth form of this species speak eloquently of the difficulty of walking though the chaparral where this species is common. An enlarged flower is shown *(lower right)*.

other chaparral shrub species. These leaves take on a darker appearance during summer and are often vertical and rolled inward on themselves. Clusters of greenish white flowers are found on the shrubs from May through July, producing fruits that ripen in early fall. This species ranges out of the chaparral into evergreen forests and redwood groves. The berries are

Plate 39. California coffeeberry takes its name from the resemblance of its fruit to that of ripening coffee. These fruits are a favorite food of wood rats.

one of the favored foods of wood rats (*Neotoma* spp.), and the bases of large old plants are prime locations for wood rat nests.

The Heath Family (Ericaceae)

The heath family is represented in the California chaparral most abundantly by species of manzanita. This genus *(Arctosaphylos)* has 60 species, and all but three grow in California. Many are regional endemics restricted to small areas in a particular part of the state, and only a few are widespread. Manzanitas predominate in chaparral from middle to upper elevations, where they are often mixed with chamise, oaks, and conifers (pl. 11, cover photograph). Mission manzanita, summer holly, and Baja California birdbush are manzanita relatives restricted to the southern part of the state and Baja California. Generally, well-known members of the heath family are found in cool, temperate locations in many parts of North America. These include azaleas and rhododendrons, as well as the cultivated blueberry.

Manzanitas

Manzanitas (*Arctostaphylos* spp.) are the most eye-catching shrubs of the chaparral. Their smooth red brown bark, sculptured trunks, and twisting stems stand out from the dull greens and grays of the other shrubs (pls. 16, 40). Individuals may be tall and statuesque or thicketlike and low growing. It is the taller individuals that generally attract attention. Although the stems are smoothly covered by red bark, the outermost layers of bark continuously peel away in papery curls. These curling pieces are usually visible along the stems.

Like chamise and ceanothus, manzanitas are major components of the chaparral statewide. About half the species occur in northern California and half in southern California, with a few such as greenleaf manzanita (*A. patula*) growing throughout the state and beyond in the Rocky Mountains. It is

Plate 40. Manzanitas are among the most easily recognized shrubs of the chaparral. On this large individual, gray areas represent portions of the stem and branches that have died, while the red areas represent those areas that are still living.

not uncommon to find them in the highest elevations where snow is a regular feature of winter and the principal source of precipitation (pl. 13). Like ceanothus, the manzanitas have evolved many species in California, and more than half the species have small natural ranges. Also like ceanothus, there are both sprouting and nonsprouting species, a trait shared only by these two genera among the shrubs of California chaparral. Manzanitas range in size from ankle-high mats to small trees, and across a wide range of elevations from sea level to openings in mountain forests. One species, the bear-berry *(A. uva-ursi),* grows all around the Northern Hemisphere from Siberia to the mountain tops in Central America, and a few other species grow across the American southwest and down into Mexico, but almost all of the others are part of chaparral.

Manzanitas are among the longest-lived chaparral species, with some resprouters perhaps 1,000 years old! Sometimes the stems have rough gray areas intermingled with areas of smooth red bark. The gray areas are part of the plant that is no longer living. Only the red areas are still alive and growing (pl. 40). This is similar to the pattern in ceanothus species, where parts of the stem die while others remain living. In manzanitas the weaving pattern is in the red stripes along the stem. Stems are often irregular in outline because of this but are not conspicuously braided in appearance as are the stems of ceanothus. The leaves are one to two inches long, elliptical, thick, and typically flat. They are often arranged vertically along the stems. The edges may be bumpy or slightly wavy, and the surfaces vary from bright shiny green to nearly white (pl. 41).

Manzanitas have delicate, urn-shaped flowers that hang upside down on branch tips. Flowers of most species are about the size of a small pea and run from cream colored to pink. The flowering season is primarily during the winter months from December through March. The fruits take months to ripen and may remain hanging on the shrubs for some time. One manzanita plant may produce 100,000 fruits

Plate 41. Big-berry manzanita showing the applelike fruits, for which it is named. The fruits turn the same red brown color as the stems when they are fully ripe.

in a season! Because fruit development takes so much energy, manzanita plants usually flower abundantly only every other year. Hanging, spiky clusters of small leaves, called bracts, and undeveloped flower buds form at the ends of the twigs in summer. These structures, called nascent inflorescences, develop for flowering the following year. A close examination of manzanita flower clusters occasionally reveals that some flowers have small holes in the side of the floral tube, near the base (fig. 16). These holes are made by short-tongued insects such as carpenter bees (*Xylocopa* spp.) that are unable to reach the nectar when approaching from the front, so they alight on the outside of the flower and chew their way in from the outside. These insects do not pollinate the flowers as do the smaller bees that enter the flower from the front, so they are called nectar robbers.

Manzanita fruits have a slightly sweet, leathery outer layer that covers a group of two to 10 hard nutlets that may be fused or separate. The common name is derived from the Spanish word for "little apples," for the abundant red brown fruits

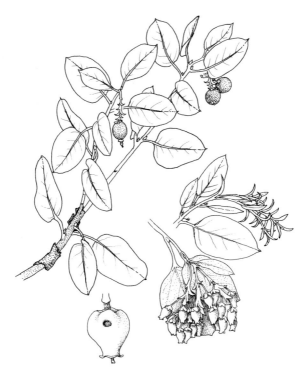

Figure 16. Foliage, flower cluster, and fruits of big-berry manzanita. This shrub has smooth, red brown bark and assumes graceful, treelike proportions. The enlarged flower (bottom center) has a hole in it made by a robber bee in search of nectar. The bee is too large to fit inside the flower, so it chews into the flower to steal the nectar, but it does not pollinate the flower.

characteristic of this genus. The fruits often ripen like apples, with the side facing the sun turning red first, while the shaded side remains green for a longer period of time. Eventually the fruits uniformly darken to the characteristic red brown color. The genus name *Arctostaphylos* means "bearberry," reflecting the importance of the manzanita fruits to both the Black Bear

(Ursus americanus) and the extinct California subspecies of the Grizzly Bear *(U. arctos)*. When manzanita fruits are in season the droppings of Black Bears, Coyotes *(Canis latrans)*, and Gray Foxes *(Urocyon cinereoargenteus)* are often made up entirely of the undigested seeds and papery outer husks of manzanita fruits. The fruits were gathered and consumed fresh by Indians, or dried and stored for later use as a source of flour and for various medicinal purposes. Indians and Spanish settlers made a beverage with the fruits, and Spaniards also used them to prepare a jelly. They were even fermented to make wine. Indians used manzanita branches as construction material, bows, and other hand implements.

There are some basic differences in appearances of the sprouting and nonsprouting manzanitas, although they are not split formally into two taxonomic sections, as with ceanothus. The leaves of sprouters and nonsprouters also do not differ markedly; both types have one main vein and hard elliptical leaves. It is necessary to look for the burl in order to confirm the ability to resprout.

Nonsprouting manzanitas, such as big-berry manzanita *(A. glauca)* of southern California, sticky manzanita *(A. viscida)* of the Sierra, and Columbia manzanita *(A. columbiana)* of northern California, have tall trunks sometimes reaching 25 feet or more in height and six to eight inches in diameter. Typically, individuals have one central trunk and rarely develop side shoots at the base. Not all species assume such large proportions, however. The widespread species Mexican manzanita *(A. pungens)* is a shrub that rarely exceeds six feet in height. In some areas these nonsprouting manzanitas form large pure stands, but they may also be found mixed with other chaparral species such as chamise or ceanothus.

Manzanitas capable of resprouting, such as shaggy bark manzanita *(A. tomentosa)* (fig. 17) from central California, Eastwood manzanita *(A. glandulosa)* (fig. 18) of southern California, and the widespread greenleaf manzanita, which grows the length of California, tend to be four to 10 feet tall,

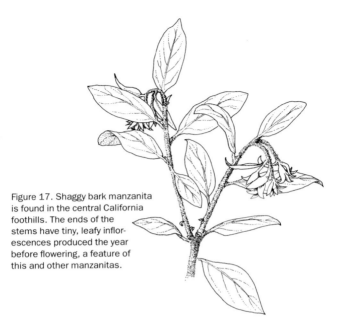

Figure 17. Shaggy bark manzanita is found in the central California foothills. The ends of the stems have tiny, leafy inflorescences produced the year before flowering, a feature of this and other manzanitas.

with numerous one to two inch stems, and are thicket forming. These species have multiple stems arising from a burl that may become a solid platform after repeated fires. These platform burls may be 25 feet across and weigh over 1,000 pounds. The plant itself is likely to be hundreds of years old. Sprouting manzanita species can form pure stands that cover many acres. As with most of the other chaparral shrubs, however, some species grow in clumps intermixed with oaks (*Quercus* spp.) or chamise or ceanothus.

A striking feature of manzanitas is the number of species that are narrow endemics. This is evidenced by species names such as Santa Cruz manzanita *(A. andersonii)*, Vine Hill manzanita *(A. densiflora)*, Little Sur manzanita *(A. edmundsii)*, Morro manzanita *(A. morroensis)*, Otay manzanita *(A. otayensis)*, Pajaro manzanita *(A. pajaroensis)*, La Purissima manzanita *(A. purissima)*, and Refugio manzanita *(A. refugioen-*

Figure 18. Foliage and fruits of Eastwood manzanita, a common burl-producing species in much of California.

sis). Each of these species is restricted to a small geographical area that is reflected in its common or scientific name. Other species that might appear to be more widespread are actually broken up into small, isolated locations. For example, while Hooker's manzanita *(A. hookeri)* grows on scattered patches of serpentine soils in central and northern California, it also includes the rarest subspecies in the genus *(A. hookeri* subsp. *ravenii),* naturally occurring only inside the Presidio of San Francisco in one small patch that is presumably a clone. Other subspecies are restricted to one canyon or site along a hillside, and several are close to extinction in the wild. The California NDDB classifies 28 taxa of manzanitas as threatened or rare throughout their range; 15 are endangered species, and one subspecies is extinct in the wild. A number of the rarer manzanitas are vulnerable to extinction because of their limited range and the fact that they are on privately held lands that might be developed.

Like ceanothus, hybridization is common between species of manzanita where ranges overlap in naturally occurring populations, and this may have been important in the formation of new species. Manzanita hybrids are produced artificially for landscape plantings, as well. Some rare species have also been introduced to cultivation in an effort to preserve individuals outside of their native habitat.

Mission Manzanita, Summer Holly, and Baja California Birdbush

While many species of manzanita are found throughout California, several of its chaparral relatives have only one species per genus and are restricted to the southern portion of the state. Mission manzanita, summer holly, and Baja California birdbush are similar to manzanita, but each has unique features of flower and fruit that have caused taxonomists to place them in separate genera. That there is only one species in each genus reflects their histories as part of an ancient flora that is now largely extinct; they are the sole survivors of what were once larger groups.

Mission manzanita *(Xylococcus bicolor)* persists in small populations from western Riverside County south to Baja California, and on Santa Catalina Island. Scattered populations are found in the Verdugo Mountains of Los Angeles County. It bears a strong resemblance to true manzanitas, as it has similar shredding red bark. It is a much-branched shrub six to 15 feet tall. It is easily recognized by its strongly bicolored leaves, dark green on top and white on the underside, with tightly rolled margins, unlike true manzanitas (fig. 19). The rolled margins make the one to 2.5 inch leaves appear narrower than they actually are, an illusion made stronger when they become vertically oriented during the dry season.

Mission manzanita has small, white urn-shaped flowers and red brown fruits much like those of true manzanitas. It flowers in the months of December through February, and the fruits often persist through the summer. This species has a burl and resprouts well after fire. Seedlings are produced only after fire. Visitors to chaparral in San Diego County will be especially likely to encounter this species.

One subspecies of summer holly (*Comarostaphylis diversifolia* subsp. *diversifolia*) is found sporadically near the coast in southern California chaparral from Santa Barbara County south into Baja and on the Channel Islands in relict populations. It is often treelike in form and up to 15 feet tall. The leaves are a shiny green, 1.5 to three inches long and about two-thirds as wide, with a finely toothed margin and rolled-under edges. It has a white, urn-shaped flower similar to that

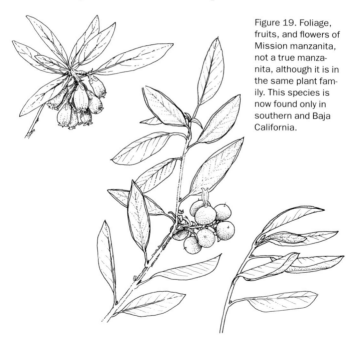

Figure 19. Foliage, fruits, and flowers of Mission manzanita, not a true manzanita, although it is in the same plant family. This species is now found only in southern and Baja California.

of manzanitas, but its fruit is a red, fleshy berry with a solid seed inside. The three largest Channel Islands have the second subspecies of summer holly *(C. diversifolia* subsp. *planifolia),* distinguished from its mainland relative by leaf edges that are not rolled under. Flowering occurs in summer after the bloom of most other chaparral shrubs has ended, and as such it is a valuable resource for insects and hummingbirds. This species does not have a burl. Many of the scattered populations of this shrub are being lost to development on private land, and it is classified by the California Native Plant Society as being endangered in part of its range.

Baja California birdbush *(Ornithostaphylos oppositifolia)* is the rarest of these manzanita relatives. It is found in San Diego County and Baja California in scattered locations, many of which are in danger due to loss of habitat as a consequence of development. It resembles summer holly in appearance, although it is generally not as tall. This species is also an important source of food for hummingbirds and insects late in the season.

Madrone, a Nonchaparral Manzanita Relative

In northern California you may occasionally come across a tall, red-barked tree with large, shiny green leaves. This is a relative of manzanita known as madrone or madroño *(Arbutus menziesii).* This is not a chaparral species, although smaller individuals may be sometimes confused with manzanitas. Madrone occasionally grows in forests or ravines adjoining chaparral areas, for example, in the San Francisco Bay Area. For the most part, madrone is common in oak forest areas and mixed with pines and other cool forest plants as far north as Washington.

Ceanothus and Manzanitas Share an Anomaly

A split in the ability to recover after fire distinguishes ceanothus and manzanitas from all other chaparral shrubs. These genera each include about 30 species that cannot resprout, and a nearly equal number that can resprout. The loss of the ability to resprout is unique to these two genera and suggests that fire has been important in their evolution.

The non-resprouting species produce no burls, and fire kills the adult. Since they cannot resprout, they are replaced in the next generation only by new plants that germinate from seeds. That they do spring forth in great numbers after fire, and only then, is proof that they are well adapted to recurring fire.

Why some manzanitas and some ceanothus, but not all, would lose the ability to resprout is not obvious. One suggestion is that more new genetic variation is produced by starting each generation from seeds rather than from resprouts from an existing burl. Resprouts are simply pruned-back parents, so there is no genetic change after each fire. New and differently adapted offspring produced from seeds are the original and genetically unique products of sexual reproduction. Over time, only those individuals with the traits most suited to the environment would survive and produce offspring, thus tracking the changes in the environment. More work remains to be done to fully understand this phenomenon. In any case, these chaparral genera, with their many resprouting and non-resprouting species, stand in obvious contrast to all the other highly successful chaparral plants—a botanical conundrum.

The proportion of non-resprouting to resprouting manzanita and ceanothus in a given area depends on a complex of environmental variables. For example, whether the percentages of the two types of shrubs will remain nearly equal among ceanothus and manzanita species, as now, depends heavily on fire frequency. This is because all chaparral plants need to grow for several years before they can produce flowers and, from them, viable seeds. For resprouting shrubs, fewer years are required to produce seeds because they can use the root system and energy in the burl to start rebuilding immediately. Still, the newly grown shoots will require a number of years before they have energy to produce a quantity of seeds. After five to seven years, resprouted branches have many leaves with which to make and store energy. At this point, with plenty of energy available, seeds are produced. The non-resprouting shrubs, on the other hand, must themselves begin life from seeds and, at least initially, grow more slowly than a resprout. This is because they do not have the physical head start of roots deep in the soil or stored energy reserves in the burl. It often takes them longer to reach the same size as a re-sprouting plant and therefore longer to flowering and seed production. Few non-resprouters produce seeds before they are five to 15 years old. Seed production in manzanita increases with age, so 40-year-old shrubs produce more seeds than 20-year-old shrubs, 70- and 100-year-old shrubs produce more than the younger ones. If fires are frequent, for example every two to three years, non-resprouters will be eliminated from that area of chaparral because they simply don't have enough time to make seeds to replace themselves before they are burned up. Resprouting shrubs may continue to resprout after repeated fires at close intervals, but even they need several years in between to recover. If the interval between fires is too short they will be weakened as they repeatedly draw down their reserves. When they fail to grow up to flowering size, no seeds will be produced to replace the parent plant. Frequent repeated fires can eventually eliminate chaparral altogether (chapter 6).

The Oak Family (Fagaceae)

Spanish colonists named chaparral for scrub oaks. The Spanish word *chaparra* is the name for a shrubby oak that grows in Spain, and its derivative "chaparral" means a place of scrub oaks. Thus, by its very name chaparral signifies the place where scrub oaks grow. This is a fitting description since they are present in nearly all chaparral. Scrub oak is often found mixed with chamise, ceanothus, and manzanita but may form extensive thickets of its own. When a stand of scrub oaks has not burned for several decades, the plants can assume the proportions of a short, compact forest. The stems become thick and twisted, with a dense overhead canopy, beneath which it is sometimes possible to walk. It was this aspect of scrub oaks and other chaparral shrubs that prompted Francis M. Fultz to title the original book about California chaparral *The Elfin-Forest of California,* a romantic title still used by admirers of chaparral. These oak thickets are among the densest patches within the chaparral and are favored by wood rats for nest building and foraging.

Scrub Oaks

California oaks all belong to the genus *Quercus.* The distinction between oak trees and scrub oaks is one of stature and form. Scrub oaks are generally less than 12 feet tall and have several to many stems emerging at ground level, while oak trees tend to have a single trunk and be up to 75 feet tall or more. In the chaparral, several species are referred to as scrub oak, the species varying with location and soil type. The most common and widespread is California scrub oak *(Q. berberidifolia)* (pl. 42). This species is found in the Coast Ranges from far northern California to Baja California, in the foothills of the Sierra, and broadly throughout southern California below 5,000 feet elevation. It may form short, dense shrubs and thickets that range from only a few feet tall to individuals with stems 15 feet or more in height. The flat- to cup-shaped evergreen leaves are .75 to 1.5 inches long. The upper surface is dark green and sometimes dusty looking, while the lower surface is covered by many short hairs and is pale green in color. The margins of the leaves are spiny. New leaves are produced in spring, gradually replacing the old leaves. The leaf litter produced by this annual turnover is deeper beneath oaks than under most other species of chaparral shrubs. Another characteristic feature found on some but not all individuals of scrub oak is the red "apple" galls (pl. 68). These bright red golf ball–sized spheres contrast strongly with the dark green foliage. These structures are caused by a reaction to insect damage to the stem and, although not a fruit at all,

Plate 42. Scrub oak, with characteristically spiny leaves, has multiple stems that may grow to several inches in diameter. Scrub oak limbs often form the framework for wood rat nests.

look very much like a ripe apple when first produced. They eventually turn brown but are woody and remain on the branches. The relationship between galls and insects is described in more detail in chapter 5.

Scrub oak flowers are produced between March and May. The brownish green tassels of the male flowers are usually conspicuous due to the large numbers of them dangling below the leaves near the ends of the branches. The small green female flowers, on the other hand, are not obvious. There are many fewer of them and the leaves usually hide them. Once pollinated, acorns, the characteristic fruit of all oaks, begin to form. The acorns have cup-shaped caps with overlapping scales that sometimes appear almost quilted. It is easy to recognize scrub oak by its leaves and by the acorn caps that persist on the branches long after the acorns have fallen. The size of the acorn crop varies between individual plants within a single population and from year to year. Since these large seeds remain on the shrubs for several months, they provide a stable source of food for many species of rodents, birds, and insects. Acorns from several species of California oak trees were a staple food for Native American cultures as well. Among all species of chaparral shrubs, scrub oaks are often the quickest to recover from fire. Plants resprout from the base and occasionally along the thick trunks within days after fire. Seed germination does not require fire, and seedlings can establish at any time. Few seedlings are seen, however, as most acorns are eaten long before the seeds have an opportunity to germinate.

While California scrub oak is the most widely distributed of the scrub oaks, several other species are also found in the chaparral. All of these are similar in appearance to the California scrub oak, making individual descriptions of each species of little value to the average person. A shrub variety of interior live oak *(Q. wislizeni* var. *fructescens)* is found in the Coast Ranges in northern and central California through to Baja California. The leather oak *(Q. durata)* is found on patches of serpentine soils in central and northern California, with one subspecies distributed in occasional places on the south slopes of the San Gabriel Mountains of southern California. Muller's oak *(Q. cornelius-mulleri)* and Tucker's oak *(Q. john-tuckeri)* are found along desert borders, in pinyon-juniper woodlands, and in chaparral in southern California. Huckleberry oak *(Q. vaccinifolia)* forms a low, dense cover in high-elevation chaparral from the central Sierra Nevada north to the Klamath Mountains and the North Coast Ranges. The Brewer oak *(Q. garryana* var. *breweri)* forms dense thickets at intermediate elevations in the Tehachapi Mountains, Klamath Mountains, southern Cascade Range, and western Sierra Nevada. Brewer oak is the only California scrub oak that is winter deciduous. Oaks, whether of the shrub or tree type, tend to hybridize readily when they grow together.

Tree Oaks

A number of close relatives of the scrub oaks are trees that are associated with chaparral. The coast live oak *(Quercus agrifolia)* is the most widespread of these. Its beautiful and familiar form is treasured by many Californians as an emblem of the natural beauty of our state. This evergreen tree is found from northern California to Baja California in valleys, slopes, and canyons, and intergrading with chaparral and woodlands. It tends to grow in places where more water is available, such as north-facing slopes, near canyon bottoms and other drainages, and areas with relatively deep soil. Coast live oaks can become very large (30 to 70 feet tall), living for several centuries and surviving many fires (pl. 43). The trunk may reach several feet in diameter and typically has gray bark that is thick and furrowed. This species has hard evergreen leaves that are somewhat larger, brighter green, and more cupped than scrub oak, with sharp points along the margins. This tree was an important source of acorns to Native Americans in many parts of the state. Several other species of oak trees are similar in appearance to the coast live oak. The most common of these are blue oak *(Q. douglasii)*, a deciduous species with a more northerly distribution that overlaps that of coast live oak and Engelmann oak *(Q. engelmannii)* in southern California. Canyon live oak *(Q. chrysolepis)*, also known as gold cup oak, is a wide ranging evergreen species that extends into Oregon and Arizona, as well as being found throughout California. Although quite variable in size, it can be distinguished from the other oaks by the gold-colored hairs on the backs of the leaves and on the acorn cups.

Another oak species with a similar distribution to gold cup oak but a very dissimilar appearance is California black oak *(Q. kelloggii)*. It favors stream edges and ravines near chaparral and extends into woodlands and into the higher-elevation coniferous forests. This oak is readily distinguished from the others by the size and shape of its leaves and its deciduous

Plate 43. The coast live oak tree is widely distributed throughout the state. It often grows in flat areas below chaparral, such as shown here, surrounded by naturalized grasses.

habit. The leaves are much larger and softer than those of the evergreen oaks, 3.5 to eight inches long and divided into wide spine-tipped lobes. In fall, the leaves turn a bright golden yellow before dropping from the tree. The conspicuous display of these yellow leaves when viewed with the blue green chaparral in the background is one of the treats of a fall trip to the mountains.

Bush Chinquapin

An oak relative, bush chinquapin *(Chrysolepis sempervirens)*, is found in high-elevation chaparral from southern Oregon to southern California. In chaparral it particularly favors rocky areas but is also found in pine forests and alpine areas up to 10,000 feet elevation. Bush chinquapin may reach eight feet in height but is generally about half that tall and grows in dense mounds and forms thick hedges along the roadsides. Bush chinquapin has a conspicuously spiny fruit that is usually

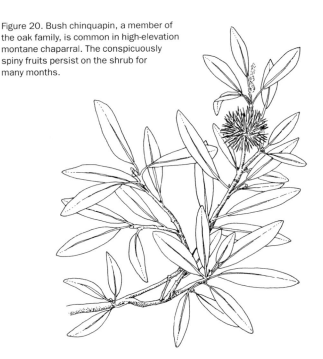

Figure 20. Bush chinquapin, a member of the oak family, is common in high-elevation montane chaparral. The conspicuously spiny fruits persist on the shrub for many months.

produced in small clusters that ripen in fall (fig. 20). The dense form of this shrub and the spiny fruits make it one of the best-defended of the higher-elevation chaparral shrubs. Bush chinquapin sprouts vigorously after fire.

The Sumac Family (Anacardiaceae)

The chaparral sumacs belong to a plant family that contains widely differing kinds of plants. For example, the sumac family includes many tropical members such as cashews and mangos and also has some highly toxic plants such as poison oak, poison ivy, and poison sumac. With the exception of poison oak, the chaparral sumacs are restricted to southern California. The typical chaparral species of this region include the laurel sumac, lemonadeberry, and sugar bush. These species

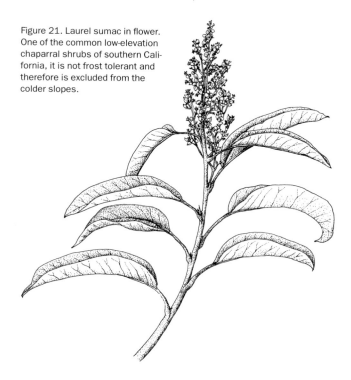

Figure 21. Laurel sumac in flower. One of the common low-elevation chaparral shrubs of southern California, it is not frost tolerant and therefore is excluded from the colder slopes.

are usually found mixed with ceanothus, manzanitas, toyon, chamise, and other species; they rarely grow in pure stands.

Laurel Sumac

Laurel sumac *(Malosma laurina)* grows on the lower mountain slopes of southern and Baja California. Typically, the plants are tall shrubs reaching to 20 feet in height. The leaves have a very pungent, distinctive odor, particularly in hot weather. The genus name, *Malosma,* refers to the applelike odor of the leaves when crushed. The leaves are large, four to 10 inches long, and have a characteristically folded appearance, rather like a taco shell (fig. 21). Unlike other chaparral shrubs, the foliage and stems of laurel sumac cannot withstand a hard freeze. When temperatures remain well below freezing all night long, as happened in southern California for several consecutive nights in the winter of 1989/90, then entire populations of laurel sumac are killed to the base. Recovery from frost damage is quick for laurel sumac, however, as plants resprout from the base of the stems just as they would do after fire (pl. 26), so that within a few years the plants will return to their original stature. The new shoots arise from various points of the root system, and not just from around the base. This pattern of resprouting is different from that of most other chaparral shrubs. In the early years of the southern California citrus industry, before reliable weather records existed for the region, pioneering farmers used laurel sumac as an indicator of places where citrus trees could safely be planted and not killed by frost. Flowers are small, about one-tenth inch across, white, and born in dense pyramidal clusters that are conspicuous in June and July. Fruits are also whitish and small with hard centers.

Lemonadeberry

Another low-elevation, near-coastal sumac is lemonadeberry *(Rhus integrifolia)* (pl. 34). This species is found below 2,000 feet in southern California, the Channel Islands, and Baja California. It rarely becomes more than four feet tall and is extremely dense and wiry. The dark green leaves are thick and leathery, one to 2.25 inches long and about half as wide, growing shiny with age. They often have a somewhat wavy surface and irregularly spaced teeth along the margins. The lower leaf surface is paler than the upper one, and during the driest times of the year the leaves become vertical in orientation. Lemonadeberry derives its name from the lemony taste of the fruit. When early California settlers mixed the fruit with water, they found that the layer of clear, sugary cells on the outside of the fruit lent a lemony tang to the water. The fruit can also be consumed fresh. Native Americans used the fruit to prepare medicines, and stems were used for a medicinal tea and for basketry. Flowers are pinkish to white in terminal clusters about three inches high. These clusters can be prominent on the shrubs in late winter and early spring because the buds are stout and often crowded. The fruits are approximately one-quarter inch in diameter, reddish, and slightly sticky.

Sugar Bush

At midelevations and interior sites from Santa Barbara to Baja California, and on the Channel Islands, lemonadeberry is replaced by sugar bush *(Rhus ovata)* (fig. 22). This shrub has multiple, stout twisting stems that may reach 15 feet tall. The

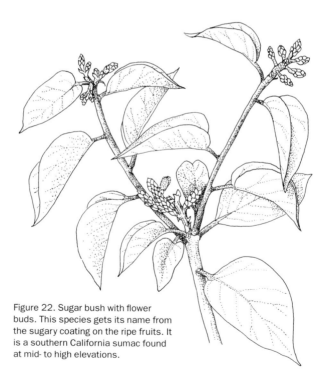

Figure 22. Sugar bush with flower buds. This species gets its name from the sugary coating on the ripe fruits. It is a southern California sumac found at mid- to high elevations.

shiny, bright green leaves are thick with a smooth margin and pointed tip. They are 1.5 to three inches long and intermediate in shape and texture between those of laurel sumac and lemonadeberry. They are often folded like those of the laurel sumac but, due to the thickness and smaller size, are more open and rounded in appearance. When crushed, the leaves of sugar bush have a pleasant odor somewhat reminiscent of apples, but distinct from the odor of laurel sumac. The year-round dark green, glossy foliage and bright pink clusters of flower buds on outer stems in late winter and early spring make this an attractive shrub that is sometimes cultivated. Sugar bush has fruits similar to those of lemonadeberry, with a covering of crystalline sugary cells, but these fruits were not generally used to make a flavored drink. Sugar bush is larger and more frost tolerant than its more coastal sumac cousins.

Poison Oak and Skunk Brush: "Leaves of Three, Let It Be"

Poison oak *(Toxicodendron diversilobum)* is so widespread in California that anyone who spends time in natural plant communities is almost sure to encounter it. In chaparral poison oak grows primarily along trails and canyons and in disturbed areas. Plants may be shrubs or vines and often lean or sprawl onto other plants. Beneath trees such as coast live oak, poison oak can become a heavy vine with thumb- to wrist-sized stems that climb into the treetops (fig. 23). Poison oak prefers cool, moist ravines, north-facing slopes, and canyons but can be found virtually anywhere in California below 5,000 feet.

It is wise to learn to recognize this plant and avoid it whenever possible. It has a three-part leaf with slightly crinkled leaflets one to four inches long. The center leaflet has a short stem and is usually the longest of the three. The new leaves that appear in spring are bright, shiny green and are much larger and softer than those of typical chaparral plants. They are also deciduous. In fall before the leaves are shed, they often

Figure 23. Poison oak growing as both a shrub and a thick-stemmed vine; foliage detail in inset. It frequently climbs into oak trees, where it may spread to cover the trunk and main branches, as shown here.

turn a bright red (pl. 44). The foliage is attractive and tempting to pick for a fall bouquet. Do not give in to the temptation! Poison oak is so named because touching any part of the plant causes a rash of itchy, oozing blisters for most people. The leaves are typically the source of the rash-causing oil (urushiol). The oil can persist on clothes, sleeping bags, pet fur, boots, skin, and any other surface with which it comes in contact. Touching a contaminated item even after a considerable period of time has passed can result in a new rash. All clothing and other items that may have come in contact with poison

Plate 44. Poison oak can be among the most attractive plants in and near the chaparral when it takes on a bright red color in late summer and fall. Do not touch this plant, however, as the oils it produces can cause a severe skin rash.

oak must be washed thoroughly as soon as possible. Exposed skin should be washed gently (so as not to introduce small cuts) and rinsed thoroughly. A dilute solution of household bleach is sometimes helpful in neutralizing the oil. Never burn poison oak, as the oil is carried in smoke and can cause severe respiratory problems.

Poison oak has a near look-alike, skunk brush or squaw bush *(Rhus trilobata)*. Both plants have soft, deciduous leaves divided into three parts. Skunk brush tends to be a lower-growing plant than poison oak, the stems are usually thinner in diameter and the leaves often hairy, but these differences are subtle and should not be used to distinguish between the two plants by anyone not thoroughly familiar with both. Poison oak has whitish flowers and seeds, while skunk brush has yellowish flowers and reddish fruits. The old warning "leaves of three, let it be" is still good advice. The berries of these species, especially those of the poison oak, are a favorite food of many birds, which spread the seeds to new locations in their droppings.

Other Chaparral Shrubs

California chaparral is characterized throughout much of its range by chamise, and species of ceanothus, manzanitas, and scrub oaks. However, outside of this group, there are other shrubs locally abundant in various parts of the chaparral. For example, coyote brush (*Baccharis pilularis,* family Asteraceae), also called chaparral broom, is occasionally encountered in and around chaparral (pl. 45, fig. 24). It is found on coastal bluffs and into oak woodlands from Oregon to northern Mexico, in the Sierra foothills and Transverse Ranges, and out to the Channel Islands. This species is highly variable, forming a prostrate mat or an erect shrub up to 15 feet high, or nearly any size and shape in between. The dark green resinous leaves are usually widest at the tip and less than an inch long, with margins that range from smooth to conspicuously notched. The flower heads are small, but usually numerous, in leafy clusters near the ends of the branches. These heads can be seen from a distance because of the protruding short, white to beige tufts of bristles. These bristles serve to disperse the seeds in the air when they are ripe. The chaparral form generally flowers during fall, but as in all other aspects of this plant, flowering time can be highly variable.

Another occasional genus of the chaparral is silk tassel (*Garrya* spp., family Garryaceae). A half dozen species are common at mid- to high elevations on the dry chaparral slopes in the coastal ranges and the foothills of the Sierra Nevada, and down into southern California. These shrubs are usually less than eight feet tall and silvery gray green with highly reflective stems and leathery vertical leaves. The leaves are in pairs on opposite sides of the stems, 1.5 to 2.5 inches long, with silky hairs that may be straight or wavy. Silk tassels can be locally abundant and are often mixed with manzanitas and other species of chaparral shrubs. Conspicuous features

Plate 45. Coyote brush, also called chaparral broom, is a common shrub in northern California. Variable in form, it may grow from ground-hugging mat to a shrub well over six feet tall.

of silk tassel shrubs are the flowers and fruits, which are produced in a dangling, tassel-like bunch from the tips of the branches. Flowering occurs in late winter or early spring, but the dry, pendulous inflorescences may persist on the plant for several months.

A showy shrub to small tree with bright yellow flowers, fremontia (*Fremontodendron californicum* and *F. mexicanum*, family Sterculiaceae), also called flannel bush, is sometimes found in chaparral. Fremontia is most common in the mountains of southern California but extends north along the Coast Ranges and the Sierra Nevada on dry granitic slopes. The foliage is rough, hairy, and sandpapery in feel. The leaves range from one to three inches long and may be wider than they are long, especially near the heart-shaped base. They may appear ruffled with deeply grooved veins. The upper surface is usually dark green, with varying numbers of hairs, while the lower surface is densely hairy and white to tawny. The flowers

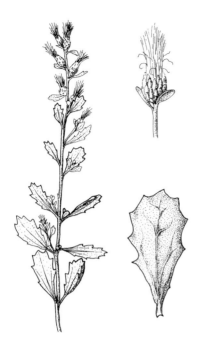

Figure 24. Coyote brush is often seen with clusters of white-tufted flowering heads along the branch tips contrasting with the green of the foliage. The enlarged leaf shows tiny resinous dots, which cover the leaf surface.

of fremontia are a gorgeous waxy yellow and often about twice the size of the leaves. Flowers may be present from March to June. The fruit has a dense brown covering of short stiff hairs that break off readily and can easily lodge in the fingers.

A deciduous plant associate of chaparral is the California buckeye (*Aesculus californica,* family Hippocastanaceae). It prefers hillsides and canyons in the Coast Ranges from near the Oregon border to Los Angeles County, as well as in the foothills of the Sierra Nevada. It varies in size from a large shrub to a small tree as much as 30 feet tall. Leaves are made up of five to seven lance-shaped leaflets, two to six inches long, each arising from the same point on the leaf stalk. Buckeyes are different from most other deciduous trees in California. They drop their leaves at the beginning of summer and remain leafless until the winter rains, instead of dropping them during winter as other trees commonly do. This is thought to be an adaptation to the summer drought characteristic of California's mediterranean climate. Near the end of spring, large, upright, white flowering stalks develop at the tips of the branches, forming a conspicuous display. The shiny brown fruit, about an inch across, is the "buckeye" that gave the plant its name. The fruits are roundish and can be seen hanging off the leafless tree branches for many months. All parts of the plant are poisonous to people. The seeds were ground up and used by Native Americans to stun fish. The flowers have also been reported to be poisonous to bees.

The chaparral currant (*Ribes malvaceum,* family Grossulariaceae) and its close relative the fuchsia-flowered gooseberry *(R. speciosum)* are two other shrub species found in chaparral (pl. 46). The chaparral currant is found primarily in the Coast Ranges throughout the state. It is three to six feet tall and has dull green, densely hairy, and roughly textured leaves, superficially resembling those of geraniums in shape. It also has pink to purple fuchsialike flowers and dark purple berries. This species blooms from fall through spring, when few other shrub flowers are available and is an important food source for hummingbirds during this time. Fuchsia-flowered gooseberry is half the height of chaparral currant, with shiny green, geranium-shaped leaves .5 to 1.5 inches long, and spiny stems. It grows in the foothills of the Coast Ranges, below 1,500 feet, from the Bay Area to Baja California. As the name implies, it produces brilliant, fuchsia-colored flowers that hang from the stems in winter and spring.

In addition to all of the species of chaparral shrubs described above, there are two other sources of diversity in the woody plants of chaparral in extreme southern California and the Channel Islands. Extreme southern California includes San Diego County and the chaparral of Baja California, which extends southward from the international border for approximately 130 miles, terminating at the southern end

Plate 46. Fuchsia-flowered gooseberry is a favorite of hummingbirds and provides nectar during winter when few other chaparral plants are in bloom.

of the Sierra San Pedro Martir. More species of shrubs per unit area occur here than anywhere else in chaparral, and many are restricted to this region. Some of the more common species are bush rue (*Cneridium dumosum*, family Rutaceae), tetracoccus (*Tetracoccus dioicus*, family Euphorbiaceae), and several species of ceanothus.

Island chaparral is found on the three largest Channel Islands. On Catalina Island, chaparral is found on the north and east slopes, while it is found on steep, rocky slopes and north-facing cliffs on Santa Rosa and Santa Cruz islands. Of the more than 600 species of higher plants found on the Channel Islands, approximately one in six is an endemic. This is the product of the islands' isolation from the rest of the state, as well as the local differences in environments. Consequently, there are unique island chaparral manzanitas and ceanothus, as well as Channel Island tree poppy (*Dendromecon harfordii*, family Papaveraceae), island redberry (*Rhamnus pirifolia*, family Rhamnaceae), and Catalina crossosoma (*Crossosoma californicum,* family Crossosomataceae), the shrub for which the journal of the Southern California Botanists is named.

Conifers: Cypresses, Pines, and Bigcone Douglas Fir

Cypresses (*Cupressus* spp.), pines (*Pinus* spp.), and bigcone Douglas fir *(Pseudotsuga macrocarpa)* are coniferous trees associated with chaparral in various localities throughout the state. Conifers produce seeds in cones and have leaves that are evergreen and scalelike or needlelike in shape. The trees have a single trunk that may reach several feet in diameter and heights of up to 150 feet. Some of these species require fire to reproduce successfully, whereas others do not (see chapter 3).

Most of the species of pines are found on the borders of chaparral, particularly at upper elevations. The gray pine *(Pinus sabiniana)* occurs on the interior slopes of the Coast Ranges and the foothills of the Central Valley (pl. 47, cover photo). Coulter pine *(P. coulteri)* is found on ridges and south-facing slopes in the mountains of southern California and the Central Coast Ranges, and knobcone pine *(P. attenuata)* grows on steep, dry sites with poor soils in central and northern California. Knobcone pine is specifically adapted to fire. Although the adult trees are killed by fire, the seeds are not released from their closed (serotinous) cones until the adults are burned.

In midelevation chaparral in the mountains of southern California, the remarkable fire-adapted bigcone Douglas fir (fig. 25) grows in dense groves nestled near the bottoms of steep-sided canyons facing away from the sun. These locations are relatively moist and lack a chaparral understory, so wildfires tend to burn beneath the trees at relatively low intensity. When fire does reach the foliage, the trees often resprout from trunks and branches that have been burned, a very unusual trait in conifers. New shoots arise from burned branches high up in the tree and occasionally along the trunk. Young trees that have never burned have the symmetrical, conical shape of an ideal Christmas tree. Older trees that have survived fires have a distinctive, open appearance, with gaps between the flat, horizontal branches. Under most circumstances, when chaparral wildfires reach a stand of bigcone Douglas fir, the intensity of the fire is reduced because the foliage of these trees does not have the high flammability of chaparral shrubs, and there is less fuel to burn beneath the trees. Bigcone Douglas fir now is thought to have a relict distribution, meaning that its present natural range is made up of small fragments of what was once a much larger and more continuous distribution. It appears that its range is still shrinking with each passing fire. This species is listed as "near threatened" on the International Union for the Conservation of Nature (IUCN) Red List.

Plate 47. Gray pine is a common member of chaparral in the foothills of central California. It is shown growing among chamise on a hillside in Mendocino County.

Figure 25. Bigcone Douglas fir, a fire-adapted tree associated with chaparral in canyons and on north slopes. The four to six inch cone is much larger than that of its cousin, the widespread Douglas fir.

Six species of cypress are present in the chaparral. A conspicuous feature of their distribution is that they are mostly confined to small, scattered populations, often upon serpentine or other poor soil types. The Cuyamaca cypress *(Cupressus stephensonii)* has the most limited distribution. It is found in only one place near Cuyamaca Peak in San Diego County. Another species with a very limited range, the Piute cypress *(C. arizonica* subsp. *nevadensis),* grows scattered among chaparral and foothill woodlands at a handful of sites in the Kern River drainages of the southern Sierra Nevada. The Tecate cypress *(C. forbesii)*, a species with serotinous cones, is confined to scattered chaparral locations from northeastern Orange County southward into Baja California. Similarly, the McNab cypress *(C. macnabiana)* and Sargent cypress *(C. sargentii)* occupy chaparral habitats in the foothills of the Sacramento Valley, and the Sargent cypress also grows on patches of serpentine soils along the length of the Central and North Coast Ranges. All have small, overlapping, scalelike leaves that cover the stems and branchlets. The cones are spherical in shape, made up of tightly fitted platelike scales that are sometimes waxy or resinous. The plates come together on the outside, giving a cone the appearance of a tiny and somewhat lumpy soccer ball. These cones are small compared to typical pine cones, typically less than 1.5 inches across.

Common Herb and Subshrub Families

To live successfully in the chaparral means coping with periodic destruction by fire. Fires burn up the shrubs whose drought and heat tolerance make them such ideal residents in a mediterranean climate. When the above-ground portions of the shrubs are removed by fire, there is a temporary change in the environment. With the shrub canopy gone, sunlight is available at the ground surface. Some of the nutrients that were formerly tied up in the bodies of the shrubs are released to the soil in the form of ashes. This clearing of the shrubs by fire, along with the nutritious ash added to the soil, allows smaller plants to dominate the chaparral for the first few years after fire. The fire cycle, a regular pattern of recovery of the chaparral after burning (discussed in chapter 3), involves a relay of plant types that replace one another in sequence until

Plate 48. Golden yarrow is one of the most common fire-following plants on chaparral burns throughout the state. It persists for several years after a fire.

the shrubs once again reclaim the landscape. In the early stages of this cycle, short (less than three feet tall), soft-bodied, short-lived plants dominate the spring displays. These are the herbs, or herbaceous plants. By the second or third year these plants are replaced by larger, semiwoody plants called subshrubs, which are two to five feet tall. The subshrubs are prominent for three to five years after fire, but they gradually give ground to the rapidly growing shrubs that overtop them, reclaiming the sun and their place as the dominant plants.

Many herbaceous plants that spring up after fire grow at no other time. Most of these are annual plants, meaning that they grow, flower, set seed, and die in one growing season. They appear only as often as fire occurs, which may be only once in a century. The shrubs and subshrubs, on the other hand, live for several to many years and are therefore perennial. Most of the over 200 species of fire-following annuals and herbaceous perennials that are found in postfire chaparral are found nowhere else.

Like the shrubs of the chaparral, many of the most obvious fire-following herbs and subshrubs belong to a small number of plant families. Some of the most common plant families on chaparral burns are the waterleaf, poppy, legume, lily, and snapdragon (figwort) families. This is far from a complete family list. There are many other plant families, some of which contribute few species, but species that may be quite important and widespread. For example, golden yarrow (pl. 48) is one of the most important subshrubs, yet few other species in the huge sunflower family (Asteraceae) are dominant in postfire chaparral at this time. The information and brief descriptions that follow provide an entrée to the wildflowers of the chaparral.

The Waterleaf Family (Hydrophyllaceae)

For sheer number of plants in postfire chaparral, phacelias (*Phacelia* spp.) surely vie for top ranking (pl. 49). They are

Plate 49. Phacelia can blanket entire hillsides the first year after fire, as shown here in the Santa Monica Mountains.

characterized by a coiled inflorescence along which are crowded cup- or bell-shaped flowers. The flowers are typically white, blue, or lavender to purple. Depending on the species, the flowers can be anywhere from one-quarter to three-quarters inch across. The stems and leaves are often bristly, hairy, and sticky. The leaves are variable in shape and size. They range from lobed and deeply divided to pinnate (with leaflets arranged laterally along a central axis like the pinnae of bird feathers, dividing the leaf into many small sections), and from .5 to five inches long. Phacelias are rarely more than three feet tall, but some climb up on burned branches or rocks and thus appear larger. They grow in dense patches that sometimes cover entire hillsides. After fire in southern California, the white flowers of the yellow-throated phacelia (*P. brachyloba*) may entirely dominate a slope face, while the blue- and purple-flowered species such as *P. grandiflora* and Parry's phacelia (*P. parryi*) are common in patches mixed with a variety of wildflowers. Caterpillar phacelia (*P. cicutaria*) can be found in disturbed areas around chaparral and in burns in the foothills of the Sierra Nevada south to Baja California (fig. 26). In northern California, the light blue phacelia (*P. heterophylla*) can be common along with the similar appearing sweet-scented phacelia (*P. suaveolens*).

Other fire-following members of the waterleaf family that might be expected after a burn in many areas of the state are species of baby blue eyes (*Nemophila heterophylla* and *N. menziesii*). These species have bowl-shaped flowers about .25 to 1.5 inches across, with dark blue or spotted petals along the outside and a pale center. The leaves are .75 to two inches long, compound, deeply dissected, and sparsely hairy. These species add to the blue and white mixtures provided by the phacelias. Whispering bells (*Emmenanthe penduliflora* var. *penduliflora*) is another waterleaf species that is common after fire (pl. 50). The pale yellow and somewhat papery bell-shaped flowers that nod along the stems give the common name to this species. The flowers are one-quarter to three-quarters inch across. The leaves are 1.5 to four inches long, sticky, hairy, and divided into numerous leaflets. This species is largely confined to the chaparral because it is dependent upon chaparral fire for seed germination.

Waterleaf species are well suited to life in the chaparral. They characteristically have sticky hairs on their stems and leaves, and the plants often are ill smelling. The combination of hairs and gummy substances produced on stems and leaves by phacelias, for example, may repel insects and other plant-eating animals. These unpleasant defenses can also be directed against people, because the hairs on some species break off and can get into the skin like fiberglass splinters. This is especially likely with dry plants because the hairs become very stiff. For the fire-specific herbs it is important that the seeds drop very near the parent plant in the chaparral, and remain

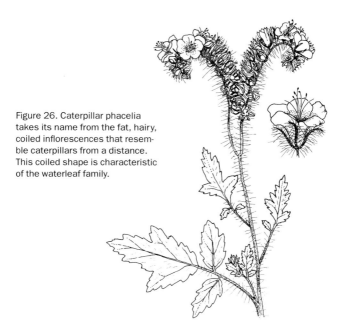

Figure 26. Caterpillar phacelia takes its name from the fat, hairy, coiled inflorescences that resemble caterpillars from a distance. This coiled shape is characteristic of the waterleaf family.

Plate 50. The pale yellow flowers of whispering bells produce seeds that require a specific chemical cue from fire in order to germinate. The seeds are very long lived, surviving in the soil 100 years or more between fires.

there. These plants are best adapted to the conditions immediately after fire, and the best place to find fire is to sit and wait right there until fire returns. The physical defenses of irritating hairs and stickiness ensure that most seeds will drop from the undisturbed parent plant in the place where they were produced. The seeds of these fire-following species are well adapted to survival in the chaparral soil between fires. They can remain viable for more than 100 years, patiently awaiting the next fire! The seeds are tiny and very dry. They contain about as much moisture as a "dry" paper towel in a typical kitchen. The small size and extreme dryness of waterleaf seeds means that they can withstand many seasons of heat, drought, or cold because they are just barely alive.

The Poppy Family (Papaveraceae)

Poppies are at home in the postfire chaparral as herbs and sub-shrubs. Some are specific fire-followers while others range more widely. The state flower, the California poppy *(Eschscholzia californica)*, is an example of a wide-ranging species. In addition to chaparral burns, this species is found in many other habitats throughout the state. These poppies add bright orange color accents on hillsides of phacelias and other fire followers (pls. 28, 51). California poppies have a rosette of leaves at the base made up of soft, finely divided leaves varying in length from approximately one to six inches. Plants are usually less than a foot tall, with many freely branched stems that tend to fall over when they become long. Individual plants may be either annuals or short-lived perennials depending on

Plate 51. California poppies and owl's clover are sometimes found together in springtime displays. These species can be found in many places bordering chaparral and in grassland areas as well.

their location. The petals are two to three inches long and a brilliant shiny orange. The petals fall off very quickly when the flowers are picked, so these plants are best admired where they grow rather than plucked and taken home.

The fire poppy *(Papaver californicum)*, unlike the California poppy, is restricted to burns, as the name would suggest. It is common after fire in central and southern California (pl. 52). The flowers are less than an inch across, brick red with green spots at the base of the petals. Flowering stems are usually one to three feet tall. Most of the finely divided leaves arise from a basal rosette, but some extend up the stem. The leaves are from 1.25 to 3.5 inches long and may be smooth or have short stiff hairs on the surface.

Cream cups *(Platystemon californicus)* is another fire-follower found widely throughout the state including the Channel Islands. Flowers are white to yellow with conspicuously crinkled petals .7 to 1.5 inches long. They are born singly at

Plate 52. Fire poppies are so named because they grow only following fire. As is typical of poppy family members, this species has a crinkled texture to the petals.

Figure 27. Prickly poppy is aptly named for its numerous spines on the stems and leaves. The petals are white with a bright yellow center of stamens. The foliage is often a silvery gray color.

the top of a slender stalk. The stems are typically hairy and can be up to one foot tall. The leaves are linear, 1.5 to 3.5 inches long, and not dissected as in the California poppy.

The prickly poppy or chilcalote *(Argemone munita)* is a larger, white-flowered poppy found on burns but is not likely to be confused with cream cups. It is called prickly poppy because it can have over 500 spines per square inch on its stems! The flowers are also larger, two to five inches, and have a bright yellow center. The stems may be several feet tall, with spiny lobed leaves two to six inches long. This species is poisonous, but because it is so spiny it is rarely eaten (fig. 27).

Another poppy genus is uniquely tied to fire in the chaparral. The common name for the genus as a whole is Dutchmen's breeches *(Dicentra)*, although this is not always used for species of the chaparral. The name refers to the pantaloon-

Plate 53. Bush poppy has some of the most brightly colored flowers of the postfire chaparral flora. This species has a special relationship with ants, enticing them to remove the ripe seeds from the bush to obtain a special reward (see chapter 5).

shaped blossoms, quite distinct from the usual open symmetrical poppy flower. Golden ear drops *(D. chrysantha),* with numerous yellow gold flowers, and its white-flowered relative *(D. ochroleuca)* are specific to the postfire chaparral. These plants can grow very tall, with flowering stalks extending six feet or more. They have waxy leaves at the base and along the stem that are pinnately dissected into many fine lobes and are from five to 12 inches long. Above the leaves, numerous pantaloon-shaped blossoms, one-half to three-quarters inch long, hang on slender branches extending from the upright stem.

Poppy family plants also grow as large subshrubs in and near the chaparral. The bush poppy *(Dendromecon rigida)* is a fire-following subshrub (pl. 53) found throughout much of the state. Bush poppy has strikingly bright yellow blossoms about two inches across, and waxy blue green foliage. The leaves are two to four inches long with smooth or finely toothed edges. The plant can reach 10 feet in height and persists for several years once established. This species is unusual because it has specialized ant-dispersed seeds, a relationship discussed more fully in chapter 5. A close relative, the rare Channel Island tree poppy *(D. harfordii)* is taller and has somewhat larger flowers than its mainland relative. Coulter's matilija poppy *(Romneya coulteri)* is not restricted to the postfire chaparral but can be found in burned areas of southern California. This poppy has the largest flowers of any native California plant, about the size of a human hand. The flower has crinkly white petals with a bright yellow center and hairy seed capsules and is three to 10 feet tall with gray green foliage. The leaves have three to five deep lobes and may be up to one foot long. Matilija poppies can be seen occasionally along highways and next to water courses. Their striking flowers make them a popular plant for parks and gardens.

All poppies contain strong chemicals to repel herbivores. A yellow or orange sap is common in the stems and roots, and it will leave a stubborn stain on hands and clothes, so be careful.

The Lily Family (Liliaceae)

The lily family includes some of the most beautiful plants of the postfire chaparral. Many species have brilliantly colored flowers, adding to the showiness of the postfire bloom (pl. 6). The flowers are often at the end of a long thin stalk that extends well above the rest of the plant. Their leaves are long and narrow with parallel veins, like grasses, and many have bulbs. Lilies are widely distributed throughout the state and are likely to be encountered on any chaparral burn. Chaparral lilies can be either small and delicate or tall and tough. The delicate ones are restricted to burns, while the tough chaparral yucca *(Yucca whipplei,* sometimes placed in the separate family Agavaceae) can be seen in openings among shrubs in mature chaparral and on very steep and rocky slopes (pl. 9). Blue dicks *(Dichelostemma capitatum)* (fig. 28) is among the most common lilies of the postfire chaparral. It is found throughout the state and north to Oregon and east to Utah and New Mexico. Plants are usually six to 10 inches tall with slender, slightly keel-shaped leaves in a cluster at the bottom of the plant. The typically blue to purple flowers are one-half to one inch long and bell shaped, growing in round heads of similar aspect to those of garden onions. Blue dicks arise from a fibrous corm, which looks something like a bulb but is actually an underground part of the stem. Since this species can grow not only in chaparral but also in many other plant communities including grasslands, open woodlands, scrub, and desert, it is likely to be encountered almost anywhere in California in springtime. Blue dicks belongs to the onion group within the lily family (sometimes separated into its own family, the Amaryllidaceae), but the corms are not edible.

Some of the most spectacular and colorful postfire followers in chaparral are the mariposa lilies (*Calochortus* spp.). The genus name means "beautiful grass," a description of the vertical grasslike leaves and the lovely flowers. Flowers of Mariposa lilies range from small, shy, nodding blossoms of white or delicate pink less than one-half inch across, all the way to open bowls two to three inches across of robust purple, blazing yellow, red, and orange. The flowers are often distinguished by the prominent colored nectar-

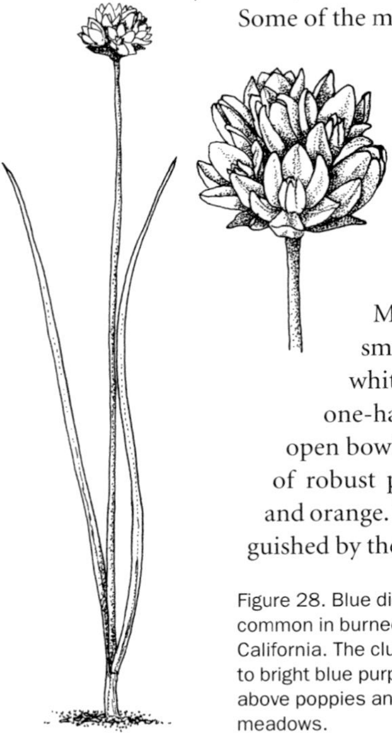

Figure 28. Blue dicks, a member of the lily family, common in burned and unburned areas throughout California. The clusters of flowers range from pale to bright blue purple and are often seen bobbing above poppies and grasses on hillsides and in meadows.

Plate 54. Splendid Mariposa lily is found throughout the state, typically in areas of recent chaparral fires.

secreting glands at the base of the petals that may also have dots or stripes on the surface (pl. 6). Over 30 species of Mariposa lilies are found throughout the state in grasslands, forests, deserts, and mountains, as well as in chaparral. As with many other California native plants, Mariposa lilies frequently have restricted distributions. The common names of a number of species reflect this: San Luis Obispo Mariposa lily *(C. simulans)*, Oakland star-tulip *(C. umbellatus)*, Tiburon Mariposa lily *(C. tiburonensis)*, and Inyo County star-tulip *(C. excavatus)*, to name a few. In the North Coast Ranges south to San Francisco, Diogenes' lantern *(C. amabilis)* is a common species of chaparral burns. Weed's Mariposa lily *(C. weedii)* is common in the South Coast, Transverse, and Peninsular Ranges. Splendid Mariposa lily *(C. splendens)* (pl. 54) can be expected in chaparral burns from the North Coast Ranges to Baja California.

Besides Mariposa lilies, star lilies *(Zigadenus* spp.), also called death camas, are common in chaparral burns. Fremont's star lily *(Z. fremontii)* and smallflower death camas

(Z. micranthus) are the species most likely to be encountered in recently burned chaparral (pl. 27). The flowering stalks may be three feet tall and stick up noticeably above many of the other herbs. Leaves are eight to 24 inches long, strap shaped, and mostly restricted to the base of the plant. Flowers are one-quarter to one-half inch wide and starlike (hence the common name) with white to greenish yellow flowers. These lilies are both widely distributed and, while not specifically requiring fire, are conspicuous on burns throughout the state. All parts of star lilies are poisonous to humans and livestock. The alternate name, death camas, reflects this. Members of the Lewis and Clark expedition became seriously ill after eating the bulbs of a *Zigadenus*.

Soap plant *(Chlorogalum pomeridianum)* is another member of the lily family that is common on burns. It is similar in general appearance to star lilies and is also widely distributed throughout the state. The one-half to one inch long flowers of this plant are unusual in that they open only at night. Closed white flowers with slender green or purple stripes are all that is visible during the day. The leaves are distinctive as well, 8 to 24 inches long with ruffled edges. The common name is derived from the soapy lather produced when juices from the bulb are added to water. Soap plant also may be found occasionally in openings in mature chaparral. The leaves are heavily browsed by deer and small mammals, so that in chaparral they are likely to mature and produce flowers only when they grow vigorously after fire.

Not all lilies are soft and bulb forming. Chaparral yucca is also known as Spanish bayonet and Our Lord's candle. It is a conspicuous member of both burned and mature chaparral in southern California. The second common name refers to the swordlike leaves, while the third refers to the tall flowering stalks often visible at great distances (pls. 9, 55, fig. 29). Yuccas grow for years in the mature chaparral but are not shrubs in the same sense and form as chamise *(Adenostoma fasciculatum),* ceanothus *(Ceanothus* spp.), or manzanitas *(Arcto-*

Plate 55. The chaparral yucca has a tall inflorescence of many creamy white flowers that can be seen sticking up above the surrounding vegetation. This is the source of another common name for this plant: Our Lord's candle.

staphylus spp.). The formidable leaves are stiff and gray green with sharply pointed tips. They range in length from one to three feet. All the leaves are of similar size and grow from a point deep in the center of the plant. Typically there is only one rosette of leaves, with leaves arising from a common center, but occasional individuals have several rosettes. After a chaparral fire it is common to see heavily singed plants with the leaves trimmed back to a central core, looking like a large charred pineapple. Flowering in chaparral yuccas is common after fire and is often spectacular for the sheer numbers. When one yucca blooms so too do the others. No one knows what makes these plants flower synchronously over large areas. These flowering stalks can be over 25 feet in height, and each stalk will have hundreds of waxy, white, bell-shaped flowers (pl. 55). Especially after fire, the tall flowering stalks are often visible for miles across the hills. A chaparral yucca plant typically has a single rosette of leaves that lives until it

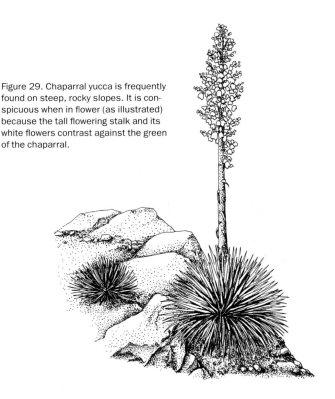

Figure 29. Chaparral yucca is frequently found on steep, rocky slopes. It is conspicuous when in flower (as illustrated) because the tall flowering stalk and its white flowers contrast against the green of the chaparral.

has finished flowering. Then the entire rosette dies, and new plants must come up from seeds. The number and size of rosettes varies with geographical location, and these variations have been recognized by some as different subspecies (see *The Jepson Manual*, in the supplemental reading section). The chaparral yucca has an interesting and essential relationship with its pollinator, the Yucca Moth *(Tegeticula maculata)*, without which neither would survive (see chapter 5).

The Legume Family (Fabaceae)

Many California plant communities have legume (bean and pea) family members as common constituents. Legumes are frequent in the postfire chaparral, where these species grow as herbs and subshrubs. Some have specific types of delayed germination caused by an extremely hard seed coat and may require fire's cue to germinate.

Deerweed *(Lotus scoparius)*, also called California broom, is one of the most common subshrubs in chaparral burns across much of the state (pl. 56, fig. 2). This plant has soft nutritious foliage and is a much-prized food item of deer and other grazers. The flowers are small, usually less than one-half inch long, and of the typical pea or bean type. Deerweed flowers are scattered among the leaves along the green stems (fig. 2 inset). The flowers are characteristically a yellow orange that turns more reddish as the flowers age. This subshrub is three to four feet tall with many soft leafy stems extending out from a woody base. The green stems as well as the leaves are photosynthetic; the one-quarter to one-half inch leaflets are often shed during the driest times of the year. Like many other

Plate 56. Deerweed, a member of the legume family, is one of the most common and widespread subshrubs after fire in chaparral. The foliage is nutritious and much favored by deer, the source of its common name.

legumes, these plants fix nitrogen in their root nodules, so they serve to enrich the soil where they grow. Over 40 species of *Lotus* are found throughout the state. These range from annuals to large subshrubs.

Lupines (*Lupinus* spp.) are also found widely in chaparral burns and elsewhere and are easy to recognize because their columnar flowering stalks look much the same regardless of scientific distinctions between species. The little spires of flowers are colorful landmarks of spring in many places throughout California. There is considerable variation in the particular species of lupine encountered in any given burned chaparral area. An example of a fire-following lupine is the southern California stinging lupine *(L. hirsutissimus)* (fig. 30). It is spectacular for its postfire blooms and defenses. The flowering stalks may be up to a foot tall, with the flowers arranged in a dense vertical column. The flowers themselves are blue purple with a yellow spot on the banner (the upraised petal above the keel-shaped lower portion of the flower). They are similar in shape to that of the Spanish broom *(Spartium junceum)*, another member of the pea family, illustrated in fig. 31. In this lupine, like many others, the yellow spot on the flower changes to a dark magenta after the flower is polli-

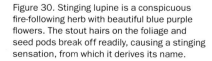

Figure 30. Stinging lupine is a conspicuous fire-following herb with beautiful blue purple flowers. The stout hairs on the foliage and seed pods break off readily, causing a stinging sensation, from which it derives its name.

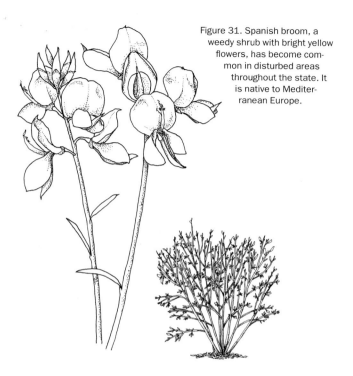

Figure 31. Spanish broom, a weedy shrub with bright yellow flowers, has become common in disturbed areas throughout the state. It is native to Mediterranean Europe.

nated. This gives bees and other pollinators the information ahead of time not to bother with this flower, since another has already taken the nectar and pollen. The leaves of this and all lupines are shaped like a human hand with many fingerlike segments radiating out from the "palm." The leaves are highly variable in size, ranging from the size of a dime to more than three inches across. The leaves may be clustered near the base of the plant in a rosette and may also extend up the stem. Stinging lupine gets its name from the stiff bristles all along the stems, leaves, and the seedpods. As the plants dry, the hairs harden. They are capable of poking through cloth with ease and may even work through a pair of leather gloves! Not all lupines are stiff and hairy. Arroyo lupine *(L. succulentus),* for

Plate 57. Arroyo lupine prominently displays the status of the individual flowers to would-be pollinators. Before pollination, the upraised banner petal is white, and after pollination a reddish spot appears. The red color can be seen on the uppermost flowers

example, has smooth, succulent stems and soft leaves (pl. 57). The leaf is usually about 2.5 to three inches wide and has seven to 10 leaflets. This species can reach four feet or more in height. Arroyo lupine is found throughout the state in disturbed areas such as roadsides, as well as in postfire chaparral. Although many lupines have blue or purple flowers, some species are yellow or white flowered.

In addition to these larger legumes, smaller plants such as chaparral sweet peas (*Lathyrus* spp.) and occasional loco weeds (*Astragalus* spp.) may also be found in recently burned chaparral. These have a typical flower for the family, the same as seen in lotus and lupines, although of different sizes and colors. Lupines and loco weeds are conspicuous from the valley grasslands and foothills into the highest mountain sites of the Sierra. As a group they are highly adaptable, as their wide elevational and geographical range attests.

Several introduced shrubby legumes, especially Spanish broom (fig. 31), Scotch broom *(Cytisus scoparius)*, and French broom *(Genista monspessulana)* are found widely throughout the state. The brooms are classified as noxious weeds by the California Department of Food and Agriculture because they have taken over many chaparral areas, especially in disturbed habitats, and are excluding the native plant species. See chapter 6 for more about the effects of these invasive species.

The Snapdragon or Figwort Family (Scrophulariaceae)

There are several fire annuals and some subshrubs among the snapdragon family in the chaparral. These include wild snapdragons (*Antirrhinum* spp.), beardtongues (*Penstemon* spp.), monkey flowers (*Mimulus* spp.), Indian paint brushes (*Castilleja* spp.), and owl's clovers (*Orthocarpus* spp.) (pl. 51), among others. These contribute much to the beauty and variety of the chaparral.

Like their cultivated relatives, native snapdragons have showy spikes of two-lipped flowers on upright stalks. The white, blue, or purple to reddish flowers of these species are smaller than those of cultivated species but are very beautiful. Fire-following snapdragons are annuals or biennials (plants that live through two growing seasons). They are usually less than two feet tall and can be upright, or twining over other plants. In chaparral, several snapdragon species are specific to burned sites, such as the viney, lavender-flowered Kellogg's snapdragon *(Antirrhinum kelloggii)*. This species is common in burned areas from northern California to Baja California. On the other hand, the twining Coulter's snapdragon *(A. coulterianum),* with white to lavender flowers, is strictly confined to southern California and Baja California burns. The many-flowered snapdragon *(A. multiflorum)* can be found in the foothills of the Sierra Nevada and from San

Francisco down the Coast Ranges and out to the Channel Islands. It has light pink to red flowers with a light brown withered area on the lower lip (pl. 58). Flowers are one-half to three-quarters inch long with the lower lip swollen and closing the throat. Although the flowers are attractive, the foliage can be hairy, sticky, and carry an unpleasant scent. Leaves grow in pairs opposite each other and are usually more numerous near the base of the stems. They may be smooth or sticky and range from .5 to 2.5 inches long. These unattractive features serve to discourage would-be herbivores. Seed capsules are oval, one-quarter to one-half inch long, and frequently sticky or hairy. The seeds are held in the capsules until shaken lose by the wind, scattering them across the soil.

Beardtongues (*Penstemon* spp.), also called penstemons, are found in chaparral before and after fire. These species are all perennial subshrubs, usually one to three feet tall. One of the most common species is the scarlet bugler *(P. centranthifolius)*. This beardtongue is conspicuous in spring and early summer because of the stalks of bright red tubular flowers that extend above the leafy base of the plant (pl. 9). Flowers range from .75 to 1.25 inches in length. The plant may reach up to four feet tall with smooth leaves and stems, occasionally with a white waxy covering. Scarlet bugler is found widely throughout the Coast Ranges and the foothills of the Sierra Nevada. In southern California this species often hybridizes with a purple-flowered relative with much larger flowers, whose scientific name, *P. spectabilis,* translates to "spectacular penstemon" (pl. 59). Over 60 species and subspecies of beardtongue occur throughout the state, and they are in many habitats outside of the chaparral, particularly in the mountains.

Monkey flowers (*Mimulus* spp.) can be common after fire and in openings in the chaparral. There are as many species of monkey flowers as there are of beardtongues, and both genera are found in many parts of the state. Species may be annuals only a few inches high to subshrubs several feet tall. A common and widely found monkey flower is *M. aurantiacus.* In-

Plate 58. Many-flowered snapdragon is common on chaparral burns the first few years after fire. This plant may grow up to three feet tall and can be either annual or perennial.

dividuals have strongly two-lipped white, yellow, orange, or red flowers, two to three inches long. The leaves are about one to two inches long, dark green with deeply impressed veins, and are sticky to the touch. It is a subshrub with numerous stems that may reach four feet tall (fig. 32). This species has many regional variants that hybridize throughout California. One curious characteristic of this plant is that the stigma, the part of female reproductive structure receptive to pollen, changes position when touched. The stigma lobes are open and outspread until a pollinator (or person) touches them. In nature, the bee or other pollinator would deposit pollen

Plate 59. Spectacular penstemon is one of the most beautiful subshrubs in chaparral.

grains on the stigma. The stigma closes up permanently after pollination, and the process of fertilization and seed development begins, secure inside the closed stigma lobes. The same effect, although temporary, may be achieved by lightly touching the stigma with a pen tip or twig. The stigma lobes will open after awhile, however, since no pollination has occurred. Flowers with closed stigmas advertise their status to pollinators who can then avoid that blossom and head for another (pl. 60). Several species of annual monkey flowers are found on chaparral burns. Some examples of fire-following annuals are the maroon-flowered species, Bolander's monkey flower *(M. bolanderi)* of northern and central California, and sticky monkey flower *(M. viscidus)* in burns and open areas in the foothills of the Sierra. A common yellow-flowered species, the short-lobed monkey flower *(M. brevipes)* is found in postfire southern California chaparral.

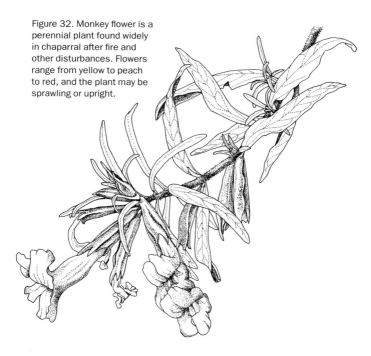

Figure 32. Monkey flower is a perennial plant found widely in chaparral after fire and other disturbances. Flowers range from yellow to peach to red, and the plant may be sprawling or upright.

Another genus in the snapdragon family, Indian paint brushes *(Castilleja* spp.) also can be found in chaparral. The common name refers to the bright red orange coloration and flared shape of the inflorescences. The true flowers are rather inconspicuous and are found down among the brightly colored leaves that make up the attractive portion of the inflorescence. A common Indian paint brush species in many parts of the state is woolly Indian paintbrush *(C. foliolosa),* a subshrub up to two feet tall found in both burned and unburned chaparral (pl. 61). As suggested by the common name, the stems and leaves are covered with white to grayish hairs. Owl's clovers are relatives of the Indian paint brushes, usually associated with grasslands but also found in chaparral burns, especially in southern California (pl. 51). Indian paint brushes

Plate 60. A close up of monkey flower shows the flattened stigma in the center. When a bee or other insect pollinates this flower, the stigma lobes close up preventing further pollination attempts.

Plate 61. Indian paint-brush, a member of the snap-dragon family, is perennial and occurs in chaparral openings after fire.

and owl's clovers cannot grow independently. They parasitize the roots of other species to obtain water and nutrition.

Other Chaparral Herbs and Subshrubs

Golden yarrow *(Eriophyllum confertiflorum)* is a member of the sunflower family (Asteraceae) and is one of the most common subshrubs in postfire chaparral in many parts of California (pl. 48). Plants may grow to three feet tall, with many stems on which are slender, deeply lobed leaves, typically one to two inches long. Foliage may be markedly bicolored with a dark green upper surface and a white underside covered with a thick mat of hairs. It has small, yellow daisylike flowers that grow in flat clusters of up to 30 heads at the top of leafy stems. A close relative, woolly yarrow *(E. lanatum)*, is also widely distributed in burns and hybridizes with golden yarrow. This latter species generally has fewer than five heads in a cluster and is densely hairy. There are many varieties of these two species, however, so that it may be difficult to separate them from each other. Everlastings *(Gnaphalium* spp.) are other sunflower family members that can be common annual plants on burns in some areas. These are short, gray white woolly annuals whose flowers are so inconspicuous that they appear to be lacking altogether. The plants may be only a few inches tall, and they have a single stem.

California buckwheat (*Eriogonum fasciculatum,* family Polygonaceae) is found widely in dry areas of coastal central and southern California and east of the Sierra Nevada to the desert (fig. 33). In chaparral it can be found scattered or in localized patches on ridges, dry washes, and exposed sites. It is often most conspicuous where road building, frequent fires, fuel breaks, and other disturbances have created inroads into the chaparral stands, particularly in the southern part of the state. California buckwheat is a multistemmed subshrub typically growing three to four feet tall but may reach greater heights under particularly favorable conditions. The leaves resemble those of chamise. They are typically less than an inch long, narrow, and produced in fascicles along the stems. However, unlike chamise, buckwheat leaves have a pale undersurface, and they are summer-deciduous rather than evergreen. California buckwheat has many, tiny pinkish white flowers (one-tenth to two-tenths of an inch across) in dense flat clusters about the size of a hand. This species typically flowers during spring. When the flowers fade they turn a rusty reddish brown, again similar to those of chamise. The dried flowers remain on the plant for much of the year. Leaves are shed during the summer drought. The flowers are a favorite of honeybees and other insects during the spring bloom, and the seeds are much sought after by animals. The tiny seeds were

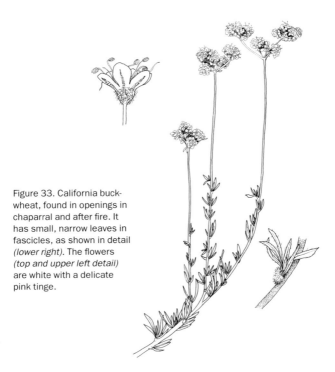

Figure 33. California buckwheat, found in openings in chaparral and after fire. It has small, narrow leaves in fascicles, as shown in detail *(lower right)*. The flowers *(top and upper left detail)* are white with a delicate pink tinge.

gathered for food by Native Americans and by American pioneers, who ground them to make flour, hence the common name.

Not properly a shrub or subshrub, wild cucumbers (*Marah* spp.) are members of the squash family (Cucurbitaceae) and are found widely throughout the state. Four species of wild cucumbers, also called man-roots, occur in the lower elevations of chaparral and associated plant communities. Wild cucumbers have a typical squash plant form, with sprawling stems; large, lobed leaves; and multiple tendrils (fig. 34). The bright green leaves one to six inches across make these plants easy to recognize. Both before and after fire, wild cucumbers are obvious in spring as they sprawl over the tops of shrubs or cover entire burned hillsides. The fruits, produced in late spring, are typically spiny and oval or round (fig. 34). The fruits eventually dry, releasing the seeds as the aboveground portion of the plant withers away. Although the twining stems may become very long, the bulk of the wild cucumber plant is below ground in the storage tuber. It is the manlike shape of these tubers that gives rise to the other common name. These tubers may reach enormous size. One such tuber, excavated with a backhoe at Rancho Santa Ana Botanic Gardens in Claremont, California, weighed 600 pounds. The tubers contain a poison used by Native Americans to stun fish. The seeds are also poisonous.

Southern California chaparral has more species of plants than chaparral of other areas and includes elements from other community types, especially coastal sage scrub (see chapter 1). For example, sage family (Lamiaceae) subshrubs are sometimes common. White sage *(Salvia apiana),* black sage *(S. mellifera),* and purple sage *(S. leucophylla)* are found at the lower edge of chaparral and in chaparral openings. In some places in southern California where chaparral has been frequently disturbed, for example, by fire at close intervals, sage plants move in as the chaparral shrubs die out. Sage species have an almost overpowering, pungent smell that per-

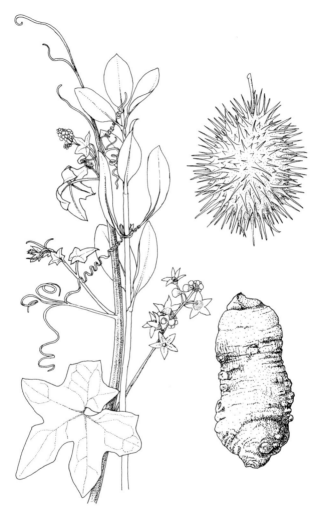

Figure 34. Wild cucumber, or man-root, a member of the squash family, is a perennial vine with an enormous underground storage stem that may weigh several hundred pounds *(lower right)*. The fruits are conspicuously spiny *(upper right)*.

vades the air around them for some distance, especially in hot weather. These sages are generally not considered true chaparral plants because they are drought-deciduous subshrubs to shrubs rather than being woody, evergreen, sclerophyllous shrubs. In low-elevation areas these plants were once found in abundance, defining their own plant community, the coastal sage scrub. This plant community has been eliminated or greatly reduced by urbanization over most of its original range, so that it and some of the individual species it contains are classified as threatened or endangered. These plants do not occur in northern California. Some other plants that may be herbs or subshrubs associated with the coastal sage and chaparral in southern California are the bush mallows *(Malacothamnus densiflorus* and *M. fasciculatus)* (fig. 35) and globe mallows (such as *Sidalcea hickmanii*) in the cotton family (Malvaceae).

An easily recognized family, also best represented in the southern part of the state, is the cactus family (Cactaceae). Cacti are part of chaparral and associated low-elevation plant communities from Santa Barbara County southward and on the Channel Islands. Cacti possess spines and succulent stems with a green photosynthetic surface in place of leaves. Most chaparral cacti belong to a single genus, *Opuntia.* Species of this genus have two distinctly different growth forms, those with flattened stems, the prickly pears, and those with cylindrical stems, the chollas (pronounced choy-yuh). The prickly pears' stems are called pads. They are the size of small dinner plates and very spiny. The prickly pear fruits are produced along the edges of these pads. The fruits also have spines, hence the common name, requiring some care in getting to the sweet edible pulp inside. The most commonly seen species on the disturbed edges of chaparral is the coast prickly pear *(O. littoralis),* but deeper in the chaparral, the much smaller beavertail cactus *(O. basilaris)* is more frequent. The valley cholla *(O. parryi)* has much narrower stem segments and more obvious branches. While the prickly pear stem segments are tightly attached, the segments of valley cholla de-

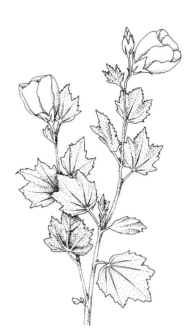

Figure 35. Bush mallow, a member of the cotton or hibiscus family, is found in openings and after fire in chaparral. The light pink blossoms are often spaced out regularly along the softly sprawling stems.

tach readily when touched. The spines catch on clothing or fur and do not dislodge readily. This is a dispersal mechanism because it ensures stems will be carried to a new location by the browsing animal or careless human that bumped into them. Detached stem segments can take root and grow when they land in a suitable site. The smoothest part of the spines of *Opuntia* species is at the tip. Below the tip they are barbed in the same direction as the spine is going on the way in. As a consequence, the spines go in easily but, like a fish hook, are difficult to remove. Caution is advised when extracting spines from your person or your shoes and clothing, as they can break off, leaving bits of the spine behind where they will likely fester for some time. Two other cactus types are sometimes seen, especially near the coast in San Diego County and Baja California. These are barrel cacti (*Ferrocactus* spp.) and fish-hook cacti (*Mammillaria* spp.).

Introduced Weeds

Some of the herbaceous plants of the chaparral are recent arrivals to California. These plants are associated with human alterations of the landscape. Roads, clearings, and fires, especially at close intervals, tend to enhance the ability of many of these herbs to gain a foothold in chaparral areas. Most of these immigrants evolved in the Mediterranean Basin and Asia Minor and were transported to California with the European colonists in the late eighteenth century. They hitchhiked as seeds in the feed, fur, and droppings of domestic animals or mixed with crop seeds such as wheat and rye. These plants include most of the annual grasses common to disturbed areas in and around chaparral, such as the bromegrasses (*Bromus* spp.) and wild oats (*Avena* spp.). These grasses, plus ryegrasses (*Lolium* spp.), are commonly seen on burns, as well as intermixed with chaparral in openings and on trails (see chapter 6). Native California grasses are bunchgrasses, which have many stems growing from a perennial base, unlike the single-stemmed annual grasses from Europe. Native bunchgrasses (*Achnatherum* spp.) have been entirely replaced almost everywhere in California by the introduced grasses (pl. 43). Two genera of introduced broad-leaved weeds, filarees (*Erodium* spp.) and mustards and black mustards (*Brassica* spp.), are common throughout the state in and out of chaparral, especially in frequently disturbed areas. Black mustards can, and often do, dominate entire hillsides throughout the state in spring, turning them a bright yellow. Fire annuals and other native species may be entirely eliminated by mustards. Collectively these naturalized annual plants from the Mediterranean are a local example of a troubling global phenomenon, where a small number of invasive plant species are displacing a larger number of endemic species. See chapter 6 for more information about the disruptive effect of these naturalized plants on chaparral.

ENTER CHAPARRAL and stand quietly for a moment. You are almost certain to hear the insistent, repetitive trill of a Wrentit, and the scratching sound of a California Towhee rummaging through the leaf litter in search of invertebrates. A Western Whiptail lizard may catch your eye with its jittery movements as it hurries from bush to bush, while a motionless Coast Horned Lizard will be noticed only when its tongue darts out to capture a harvester ant. A careful eye can find the footprints of a kangaroo rat and perhaps the entrance to its burrow. Examination of the base of large old shrubs may reveal the impressive stick nest of a wood rat. Perhaps one of its more interesting collections, such as a skull of another animal, can be seen in among the sticks. Walking farther requires watching for the Western Rattlesnake, which might be unobtrusively curled up anywhere. In warm weather, flies buzz above the shrubs and are snapped up by the sallying Ash-throated Flycatcher. After a winter rain you might encounter an ambling Ensatina Salamander that has just emerged from its basement-level shelter in a wood rat's nest. Visit at dusk and you will hear the stacatto tail drumming of wood rats, announcing their presence to one another. Just after daylight, near the edges of trails and roads, you are apt to see deer browsing with deliberation. If you are very lucky you may spot a Bobcat or Mountain Lion hunting for its dinner. There is always abundant evidence of chaparral's animal inhabitants. Most species are present year-round, and their activities are an essential part of the functioning of chaparral ecosystems. Encounters with animals are an important ingredient in the complex flavor of the chaparral experience.

Like the plants, the animals of chaparral must live with the rigors of a mediterranean climate. Temperature extremes and the scarcity of water for much of the year are the major factors that determine activity patterns and lifestyles. As a consequence, most mammals are nocturnal or crepuscular (active at dawn and dusk) because these are the coolest parts of the 24-hour daily cycle. These animals spend their days in underground burrows, rock crevices, or insulated nests that protect them from the high daytime temperatures. Reptiles, some birds, and many insects active during the day typically control their temperature by alternating between sunlit areas and the shady branches or cool soil beneath the shrub canopy. In winter when temperatures are too cold, some of these animals hibernate or otherwise become dormant. Seasonal changes and fire provide additional variations and resources, as do the different elevations and the varying topography on which chaparral is found statewide. The chaparral as a whole contains more species of mammals, birds, and reptiles than most other ecosystems in California because of these features.

Animals are unlike plants in that they move from one place to another to achieve a comfortable environment. For the chaparral, this means that many animals with good mobility and a range in diet may be present at some times and not at others. Most can live in other habitats. For example, wide-ranging animals such as Mule Deer, California Towhees, Coyotes, and Western Rattlesnakes live in chaparral, as well as in forested areas, grasslands, and many other habitats. However, a relatively smaller number of animals is particular only to chaparral. In this chapter we focus first on those animals that are chaparral specific and whose biology is known. For example, wood rats, Wrentits, Pacific Kangaroo Rats, California Whipsnakes, and harvester ants are particular to the chaparral. Among these are animals with intriguing habits and special adaptations to chaparral. The amount of information for mammals, birds, reptiles, and insects is uneven, however, as are their numbers. Some of the best stories of adaptation to chaparral are found among the insects, yet overall, this is a poorly studied category. Besides those animals particular to chaparral, we also include the more common wide-ranging animals that frequent chaparral and may be encountered by a visitor. Chaparral animals often avoid the harsh glare of open sunlight, and many do not leave the protective cover of the shrubs by day, or are entirely nocturnal. Seeing these multi-talented inhabitants takes some patience and care.

Mammals

Thirty-eight species of terrestrial mammals, plus about a dozen species of bats, regularly occur in chaparral. This is approximately 25 percent of all those found in California. Of these, 11 rodents and one rabbit are endemic to chaparral. Rodents are the most common animals in the chaparral throughout the state. Most are directly dependent on the chaparral plants for food and nesting sites. A number have specialized diets of seeds and leaves. Other herbivores such as rabbits, squirrels and chipmunks, deer, and Bighorn Sheep are more general in their habitat and eat a wide range of plants. Carnivores such as foxes, Mountain Lions, Bobcats, and Coyotes are also generalists. These animals use the chaparral for hunting, as they do forests, deserts, and other habitats.

Rodents (Order Rodentia)

Approximately 21 species of rodents inhabit the chaparral. Of these the most common are the wood rats, kangaroo rats, and white-footed mice. Typically, these animals feed on seeds, foliage, and insects. They are small and agile, and many prefer to move along the stems of the shrubs rather than on the ground. They have acute hearing, excellent vision, and a good sense of smell. They are also good at moderating their temperature and water loss through a combination of behavior, nest construction, and physiology. To avoid overheating and water loss, as well as predators, they are active at night (nocturnal) or at dawn and dusk (crepuscular). Either end of the day is a good time to look for these animals, and it is also the time of day when the diurnal (day-active) birds, reptiles, and insects cross paths with more nocturnal species, like factory workers changing shifts. Usually only a few species of rodents (two to eight) are found in any one area of chaparral. Some are highly territorial and will defend their feeding and nesting areas against other members of the same species. The particular species present and their relative numbers change markedly with the age of the chaparral relative to the last fire.

Wood Rats

Wood rats (*Neotoma* spp., family Muridae) are the signature animal of mature chaparral throughout the state. These nocturnal rodents, also known as pack rats or trade rats, may weigh up to two pounds and are the largest of the chaparral species. They have large ears, bright eyes, and a pleasantly intelligent look about the face. They are gray brown and rather chunky looking for a rodent, with a body eight to 10 inches long and a tail of equal length (fig. 36). The Dusky-footed Wood Rat *(N. fuscipes)* occurs in northern California chaparral, while the Big-eared Wood Rat *(N. macrotis)* occupies chaparral in southern California. Wood rat numbers increase as chaparral shrubs age and increase in size. Their favored habi-

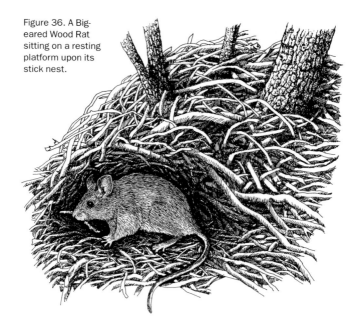

Figure 36. A Big-eared Wood Rat sitting on a resting platform upon its stick nest.

tat is among scrub oaks (*Quercus* spp.), where the solid horizontal limbs serve as pathways to and from foraging areas and nests, which may also be in among the scrub oak branches. These rats prefer to travel along branches rather than on the ground, where the rattle of leaves might give them away to would-be predators. Since a wood rat is a relatively heavy rodent, this branch-traveling behavior is only possible where the shrubs have grown large. Scrub oaks, toyon (*Heteromeles arbutifolia*), California coffeeberry (*Rhamnus californica*), and other more treelike members of the chaparral are their preferred habitat. Since these shrubs are often found in ravines or on moister north- and east-facing slopes, these are likely places to find wood rats. They are nocturnal and can see and hear well in the dark, and they also have an excellent sense of smell.

Wood rats are notable for their large and complex nests made of plant stems and twigs. The nest provides protection from changes in temperature and humidity, as well as from predators. A wood rat adjusts its environment through the architecture and location of its dwelling. On a summer day, temperatures inside a wood rat nest may be 15 to 30 degrees F lower than on the nearby exposed soil surface, and it will stay comfortably warm on a cold winter's night. Because heat is the greatest problem for most of the year, the nests are usually located in cool spots such as north- or east-facing slopes, or in a ravine or rock cave.

A nest is shaped like a small haystack and can be up to five feet high and 10 feet wide (fig. 36). Its construction is by no means haphazard despite its jumbled appearance. Twigs, leaves, branches, and other plant parts are firmly woven together to create strong walls on the outside. Inside are numerous chambers, trails, verandas, food caches, and more. Large nests may contain several living and sleeping areas, a larder, and exterior platforms used as latrines. In addition to its aboveground rooms, the nest incorporates tunnels and ground chambers for protection and escape. The nests are surprisingly roomy. If the ratio between the volume of a wood rat's body and the volume of its nest were applied to human dwelling space, a person would have a 13 by 13 foot room, or about the floor space of a modest American bedroom plus bath. Many of the chambers and other spaces are so large that other kinds of animals use the nests as uninvited and often unnoticed guests; there's a whole ecosystem inside them (see below).

Finding food and plant materials for their nests is a daily, year-round activity for wood rats. Since the chaparral is evergreen, fresh plant food is always available. Wood rats have a broad vegetarian diet. They consume seeds, buds, flowers, and leaves and specialize on the moist, nutrient-rich cambium tissues just below the bark of the chaparral shrubs. Cambium is especially important in late summer and fall when many other plant foods are unavailable. In pursuit of this moist tissue, they gnaw bark, twigs, and branches off of the shrubs, eating some and piling others on the nest, thus maintaining its strong exterior. Parts of virtually all shrubs close to the nest will be used as food. If an animal finds something good to eat on the journey out from the nest, it will be eaten on the spot. If there is a rich supply of a preferred food such as acorns or California coffeeberries, rather than eating on the road, the food will be brought home and stored in larders, some of which can hold up to 20 pounds of food, 10 to 15 times the body weight of the animal! When eating at home a wood rat often dines on the lookout porch or terrace near the entrance to its house, where it can survey the surrounding terrain and watch for predators.

While their native food and housing materials are the plants of the chaparral, wood rats may incorporate any interesting item that comes their way. In nature these items might be small animal bones including skulls, pieces of cactus, dung of larger animals, or bits of bird nests. When people are in the area the items may be more unusual. Nests have contained surveyors stakes, broken bottles, tin cans, nails, baling wire, automobile bolts, pieces of inner tubes, newspapers, plastic bottles, sweatshirts, pliers, and miscellaneous household objects! Wood rats have the intriguing habit of leaving an item in the place of the one taken; a small stone or twig will replace a shiny button, for example. Since the rat carries objects in its mouth, if it already has an item and encounters another that it finds more attractive, then it must put the one down to pick up the other. This habit of seeming to exchange one item for another explains its other common name, the Trade Rat. There are countless stories of wristwatches, rings, coins, and the like being exchanged for twigs, rocks, and other such items. A fun but unlikely story is that of a wood rat taking a small object from a miner's camp and leaving a gold nugget in exchange.

The nests, like rental houses, are occupied over the years by a succession of tenants. It is generally safer and more efficient for a young animal setting up housekeeping to move into an empty nest rather than start an entirely new one. Each nest usually has only one adult resident at a time because these species are highly territorial, and one rat will drive others away. Wood rats often occur in dense populations in mature chaparral, reaching about 16 adults per acre. This tight packing is possible because each rat forages close to its nest and rarely runs directly into another rat's territory.

A wood rat house is a ready-made habitat for a host of other animals that find shelter and food in this cozy environment. These hangers-on and inadvertent guests use the temperature stability and physical protection of the wood rat's house to their advantage. Among the most interesting denizens of these nests are the Giant Flea, the fungus-feeding beetles, and the bot flies.

The Giant Flea (*Hystrichopsylla dippieei,* family Hystrichopsyllidae, order Siphonaptera) is a common member of Dusky-footed Wood Rat households in northern California. For a flea it is huge. The female can be .25 inches long rather than the pinhead size (less than .03 inches) of ordinary fleas.

These fleas can jump only two to three inches because they are so large. However, once inside the wood rat's living chamber there is no need for long leaps. They settle onto their rat hosts, which then supply the flea with food and water. The population of large and ordinary small fleas in a single nest can range from less than 100 to well over a 1,000 individuals.

Cryptophages (family Cryptophagidae, order Coleoptera), a specialized group of beetles that are found in wood rat houses, eat the fungi that grow thickly on decaying plant matter in the lower reaches of the nest. After a wood rat has eaten most of a twig, it may drop the leftover bit on the floor. As this garbage piles up it becomes a perfect source of food for fungi that live on decaying plant material. As the fungi eat the leftovers of wood rat dinners, they themselves become dinner for the fungus-eating beetles.

Bot flies (*Cuterebra* spp., family Cuterebridae, order Diptera) have an obligatory and peculiar relationship with chaparral wood rats. The adult bot fly lays her eggs near the entrance of the nest. After hatching, the young larva drops on a passing wood rat and quickly burrows inside the warm body. As it grows it moves just beneath the skin and forms a chamber, called a warble, which protrudes from the body. The larva grows to an inch in length, feeding on the body fluids of the rat. The center of the bulge has a breathing hole for the larva within, which can be seen squirming within through this small opening. At maturity the larva drops out of the hole onto the ground and then pupates. Grotesque and traumatic as these warbles appear, they quickly heal when the larva departs, and they seem to do no permanent harm to the host. About 50 percent of all wood rats will have one to eight bot fly larvae in their bodies during spring and early summer.

In addition to these invertebrate specialists, many kinds of spiders, ants, beetles, centipedes, and pseudoscorpions are regularly found in and around wood rat nests. There is even a mouse that lives in the nest at times (see California Mouse [*Peromyscus californicus*]). All these animals are attracted to the many resources of this special environment, which are otherwise unavailable in chaparral. Nests also often shelter Ensatina Salamanders *(Ensatina eschscholtzi)*, which otherwise could not live in most chaparral because it is too dry. Although uninvited, these houseguests seem to be accepted by wood rats without conflict.

The Desert Wood Rat (*N. lepida,* family Muridae) is found in recently burned chaparral and open chaparral, particularly on very steep and rocky slopes (pl. 62). It also occurs on all the deserts of California and the Great Basin, hence the common name. It is similar in appearance to the Dusky-footed Wood Rat and the Big-eared Wood Rat, but somewhat smaller in overall size. The Desert Wood Rat has a catholic diet of various plant parts and plant species, including the succulent tissues of prickly pear and cholla cactus (*Opuntia* spp.). Juicy cactus tissues contain oxalic acid, which in large quantities is poisonous to most mammals, but it leaves the Desert Wood Rat unaffected. Its nest is constructed from sticks, twigs, pieces of cactus, and other portable objects piled over holes or openings between rocks. Rock outcrops and patches of cactus are particularly favored. The assortment of twigs and debris over a Desert Wood Rat nest is never as extensive or well organized as the elaborate structures of the Dusky-footed and Big-eared Wood Rat. A Desert Wood Rat nest is most easily found by inspecting rocky areas and cactus patches for the odd little piles of material placed across openings to nests. If a Desert Wood Rat is living within, piles of dark feces about the size of cooked rice grains will be found on the tops of nearby rocks or boulders.

Kangaroo Rats

Another characteristic group of chaparral rodents are the kangaroo rats (*Dipodomys* spp., family Heteromyidae). They are easy to recognize with their wide triangular heads with large, round shiny black eyes stuck on the sides like onyx jewels, long hind legs and stubby front legs tucked up under

Plate 62. A Desert Wood Rat, often associated with rocks and cactus in chaparral.

their chin, and a very long dark and white tail tipped with a bushy tuft of long hairs (fig. 37). They weigh 1.7 to 3.5 ounces, and more than half of their 10.5 to 13.4 inch head and body length is due to the tail. Five species that look very much alike are found in chaparral, but each local area has only one species. The Delzura Kangaroo Rat *(D. simulans)* and the Pacific Kangaroo Rat *(D. agilis)* occupy the chaparral through Baja California and the Transverse Ranges of southern California. Heermann's Kangaroo Rat *(D. heermanni)* replaces them in the Central Coast Ranges and the foothills of the southern and central Sierra Nevada. The Narrow-faced Kangaroo Rat *(D. venustus)* is found in the chaparral and coastal scrub of the Central Coast Ranges, between northern San Luis Obispo County and the Bay Area. These species are replaced in the chaparral of northern California by the California Kangaroo Rat *(D. californicus)*. These five species are similar in size, appearance, behavior, and habitat requirements. All are nocturnal. Field studies suggest that they are ecological

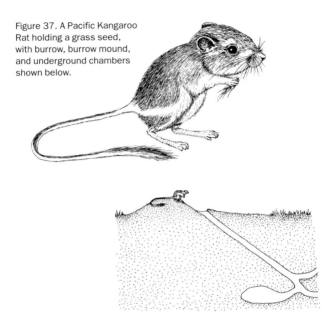

Figure 37. A Pacific Kangaroo Rat holding a grass seed, with burrow, burrow mound, and underground chambers shown below.

equivalents. That is, they have the same role in each chaparral community. Kangaroo rats are especially abundant in the first few years after a chaparral fire when large spaces between the shrubs provide clear space for hopping about, but they can be found in most chaparral with patches of open ground. Kangaroo rats do not occur in chaparral where the soil is too hard, thin, or rocky for digging, or on steep slopes.

Kangaroo rats get their name from their hopping style of locomotion that resembles that of the kangaroo. As a kangaroo rat hops along the ground covering several feet in each leap, the tail, held high in an upturned arch, switches from side to side with each bounce. This tail switching shifts the animal's center of gravity at the start of each hop, causing its path of travel to zigzag this way and that. This behavior is thought to be a defense against predators in hot pursuit, who may snap futilely at the tantalizingly conspicuous white tail tip that is in constant, erratic motion. Even if the predator snaps off the end of its tail, the kangaroo rat can do without it.

Seeds are the basis of the kangaroo rats' existence. They forage for surface and buried seeds, using an unusually acute sense of smell that enables them to find pockets of buried seeds by odor. Kangaroo rats have fur-lined cheek pouches in which they stuff seeds until the pouches are full, so that they protrude from the sides of their face as if they have giant toothaches on both sides. They return to their burrows to empty the pouches and set out again to hunt for more, making several trips each night. Kangaroo rats are strictly nocturnal, so they avoid the desiccating daytime heat and the visual predators that hunt by day.

Kangaroo rats have a unique metabolism based upon their diet of seeds. Not only do the seeds provide energy, but directly and indirectly, they also provide water. Unlike many other species of chaparral animals, kangaroo rats never drink water. Instead, their digestive system breaks down the oils and stored fats of the seeds, which releases water as well as energy. This metabolically derived water is absorbed and retained by the body. In captivity, kangaroo rats from chaparral will eat succulent plant foods to supplement water made from seeds.

Kangaroo rat burrows are very well designed to trap and conserve humidity in the air, to maintain a comfortable range of temperatures year round, and for protection against predators and wildfires. The burrows are one to two feet deep in the ground, with curving tunnels and various chambers for resting and food storage (fig. 37). The soil provides such an effective buffer from the surface environment that temperatures in the deeper portions of the burrow vary by less than 20 degrees F over the entire course of a year, regardless of summer heat and winter cold. The burrow traps air and holds it in still pockets, so that the air has a higher relative humidity than does outside air. The moisture in the burrow atmosphere is further increased when the entrance is sealed during the day. A burrow system that has existed for several generations of occupants can have multiple entrances, with a complex of tunnels and chambers that are as much as five feet across. An observer walking across the top of a burrow complex will often feel the ground under foot give way slightly as shallow tunnels are caved in. The atmosphere within kangaroo rat burrows also contributes water to their diet. Having the air in the burrow more humid than the outside air allows seeds that have been stored as food to absorb some atmospheric water. When the kangaroo rats eat those stored seeds, they gain that much more water in the bargain. Because the burrow air is comparatively humid, the amount of water lost from evaporation by respiration is also reduced.

In daylight, the presence of kangaroo rats is most easily detected by looking for burrows and for depressions from their dust baths. Burrows have openings two to four inches across, leading to a sloping tunnel. These openings may be unobstructed or closed by dirt plugs during the day. As shown in fig. 37, some burrow entrances have a slight mound on one side, where the digging rat has kicked dirt out and away from the tunnel. In places where a burrow has existed for a long time, it may lie beneath a slight mound made from the castings of generations of kangaroo rats throwing excavated dirt in all directions. Kangaroo rats must regularly take dust baths to condition their long, silky fur. These baths leave slight depressions, six to 12 inches across, covered with powdery dust. The characteristic tracks of kangaroo rats can be seen in and around these baths, as well as in other places with loose soil, as a pair of ski-shaped prints about .7 inches long, with a tail trace between. The best time to look for seed excavations, tracks, and dust baths is in early morning, when the low angle of the sun casts shadows in fresh depressions, and before the day's wind has blurred the marks of the previous night's activity.

Unlike almost all other species of small mammals, kangaroo rats often survive chaparral wildfires. They do so by remaining in their well-protected burrows. At the depth of their nest chambers, temperatures increase during a summer fire from perhaps 75 degrees F to a temperature in the 80s, which is cool enough for their survival. Dirt plugs at the entrances, plus the fact that hot air rises, prevent toxic combustion gases from entering the burrow, even if the entrance is unplugged. The aftermath of the fire is the open ground preferred by kangaroo rats, and seeds buried at depths of more than three-quarters inch remain intact as food. Consequently, fire creates and enhances habitat for these animals, so that their populations increase severalfold after a chaparral wildfire.

Kangaroo rats also have a specialized system of waste elimination that uses much less water than that of most other mammals. Their urine is concentrated to the consistency of paste, and the feces are so dry as to have the qualities of little black pebbles. Additionally, before air is exhaled from the lungs, the water in the moist breath is condensed and reabsorbed in special nasal passages. Kangaroo rats lose less than half as much water by evaporation as either white rats or humans, in proportion to their size. They can survive, under laboratory conditions, on a diet of nothing but dry seeds and salt water!

White-footed Mice

Five species of white-footed mice (*Peromyscus* spp., family Muridae) inhabit chaparral, the California Mouse *(P. californicus)*, the Brush Mouse *(P. boylii)*, the Cactus Mouse *(P. eremicus)*, the Deer Mouse *(P. maniculatus)*, and the Pinyon Mouse *(P. truei)*. As a group they are medium sized rodents, six to 11 inches long including tail, with chestnut brown backs, and bellies and feet that are white. White-footed mice have a broad diet that includes fruits, flowers, seeds, leaves, arthropods, and fungi. The five species vary in abundance according to location and the structure of the plant community, which is a function of the time since fire.

In the mature chaparral, the California Mouse is the common species of this genus. It is found in all dense chaparral from San Francisco Bay southward to Baja California in the Coast and Peninsular Ranges, and throughout the Sierra

Nevada foothills from Mariposa County southward. It prefers the same range of dense, shrubby habitats as do wood rats. The adult California Mouse is dark brown on the back, and it has a chunkier body than other white-footed mice, weighing one to two ounces (fig. 38). Besides building its own nest, it sometimes occupies the nest of a wood rat as a tolerated squatter, which explains its older common name, the Parasitic Mouse. The California Mouse and the Brush Mouse are partially arboreal, climbing among the stems and branches of shrubs with agility, using their long tail like a monkey does, as a fifth grasping appendage, and for balance. This is especially true of the California Mouse, which prefers to move about above ground. A California Mouse will climb out to the end of a branch so small that it bends under the animal's weight, and then deliberately leap across a short gap to the branch of an adjacent shrub. This species makes ball-shaped nests from twigs, generally placed high up in the shrubs. These mice form long-term pair bonds, and the parents share in care for

Figure 38. A California Mouse on a manzanita branch. Note how the tail is used for grasping the branch.

Plate 63. A Brush Mouse. The long tail, useful in climbing, is indistinctly striped and has a thin tuft of hairs at its tip.

the young and defense of the nest. The Brush Mouse looks like a slightly smaller, leaner California Mouse (pl. 63). Where either the California or Brush Mouse is common, the other species is usually rare or absent.

For a year or two following chaparral fire, the common and wide-ranging Deer Mouse is often the most numerous species of rodent in chaparral. Within its continental range this mouse often plays the role of an opportunist, making first use of the disturbed habitat created by the fire, before competing species of rodents arrive. In postfire chaparral its numbers generally decline as populations of other species of *Peromyscus,* and other rodents, increase. The Deer Mouse does not climb, and as shrub cover develops it is largely replaced by the climbing species of white-footed mice. In mature chaparral, if it is present at all, the Deer Mouse is found in whatever open spaces exist, and at edges between chaparral and other plant communities.

In southern California the Cactus Mouse is often the second species of *Peromyscus* to occur in numbers after chaparral

Plate 64. The Pinyon Mouse has unusually large, rounded ears for a white-footed mouse.

wildfire, displacing the Deer Mouse within a few years. It is somewhat smaller than the Brush Mouse, with a nose-to-tail-tip length of about six to eight inches, and a weight of .7 to one ounce. The upper body is gray brown and the long tail is thinly haired. It is most common in postfire chaparral at the stage when there is still some space between the shrubs, while herbaceous plants and seeds are common. It is also found in chaparral with scattered shrubs, and in other plant communities with shrubs and cacti, throughout the American southwest and northern Mexico. The maximum population densities of the Cactus Mouse are lower than peak densities of other white-footed mice, and sometimes more seasonally variable.

In southern California the Brush Mouse displaces the Cactus Mouse three to five years after fire. It generally prefers shrubby vegetation that is somewhat more open and dry than the California Mouse, the specialist of mature chaparral, but the two may coexist in some areas where the chaparral includes irregularities such as rock outcrops that offer a mixture of habitat types.

The Pinyon Mouse is a moderately large white-footed

mouse, with a head and body length of 3.5 to four inches, and a distinctly striped tail about the same length (pl. 64). It is most easily recognized by its large, rounded, Mickey Mouse ears that are about one inch long. The species is widely distributed throughout California in chaparral and other communities where some shrub or tree cover is available. The Pinyon Mouse does not occur in chaparral on the coastal side of the mountains of southern California, where its place seems to be taken by the Brush Mouse. In chaparral these two species seem to compete in places where their ranges overlap.

Harvest Mouse and California Pocket Mouse

Apart from the white-footed mice, two other small mice, the Harvest Mouse (*Reithrodontomys megalotis,* family Cricetidae) and the California Pocket Mouse (*Chaetodipus californicus,* family Heteromyidae) are found occasionally in the chaparral. They are particularly abundant the first few years after fire. In more mature chaparral they are usually found along its margins, or in a disturbed area with grasses intermixed with shrubs. The Harvest Mouse is found across most of California and the western United States, usually in places with thick weeds or grasses growing close to the ground. In chaparral it is occasionally found in the first few years after fire, when the ground is covered with dense herbaceous growth. The California Pocket Mouse is found primarily in chaparral and adjacent habitats, from San Francisco Bay and the central Sierra Nevada foothills southward into Baja California. The distribution of this species appears to be spotty, and its numbers are variable. When present, the California Pocket Mouse seems to prefer edges between chaparral and other plant communities.

The Harvest Mouse is the smallest rodent to be found in chaparral. It has a head and body length of three inches, and a tail of approximately equal length. The back is gray to brown, with the belly and underside of the tail white to deep gray. It looks something like a slender House Mouse *(Mus musculus).*

The California Pocket Mouse is a small relative of kangaroo rats common in chaparral after fire. It has a head and body about 3.5 inches long, with a longer tail with a tuft on the end. The back fur is a mixture of yellow and black hairs, with spiny hairs that stick out on the rump. Like kangaroo rats, it has cheek pouches, and the hind feet are relatively large, but unlike kangaroo rats, it does use both front and back limbs to move.

Squirrels and Chipmunks

Squirrels and chipmunks are rodents belonging to the family Sciuridae. They occur in chaparral but are generally more abundant in forests, woodlands, and open areas adjacent to chaparral. The California Ground Squirrel *(Spermophilus beecheyi)* is the most widespread squirrel in the state and the only species found around the edges of chaparral. It is easily distinguished from the Western Gray Squirrel *(Sciurus gresius)* and other tree squirrels because it lacks a thick bushy tail, is usually seen on open ground, and does not climb or keep near to trees. Around chaparral, the California Ground Squirrel is generally found near human disturbances such as road cuts, plowed areas, and agricultural fields. It has expanded its range into areas of chaparral where roads, trails, power line rights-of-way, and islands of urban development have dissected the previously impenetrable shrublands. It digs extensive burrows and forages in disturbed areas with relatively open ground. The California Ground Squirrel is not commensal with humans in the sense of the House Mouse and Roof Rat *(Rattus rattus)*, sharing our dwellings, but it does invade and occupy open landscapes created by human activities.

Chipmunks (*Tamias* spp.), like squirrels, are associated with openings and disturbance in chaparral and woodlands, especially those caused by people. Three species of chipmunks can be found in chaparral. Merriam's Chipmunk (*T. merriami*) occupies the chaparral of southern California through the Coast Ranges to San Francisco Bay, and the foothills of the

Figure 39. A Merriam's Chipmunk. Unless foraging on the ground, as shown here, it is more likely to be perched on a log or some other raised vantage point.

southern and central Sierra Nevada. The Sonoma Chipmunk (*T. sonomae*) replaces it in northern California. The Obscure Chipmunk (*T. obscurus*) occupies chaparral of the Transverse and Peninsular Ranges of southern California and Baja California. All chipmunks are easily recognized by their size, with a head and body length of about nine to 11 inches, a bushy tail somewhat shorter than the body, and a brownish color with distinct light and dark stripes from nose to rump (fig. 39). One of the dark lines always passes through the eye. Even the most casual observer usually notices chipmunks, because they are busy in the daytime, noisy, and often conspicuously perched on little vantage points such as rock tops, where they make a scolding chatter when approached. Unlike their cousins, the kangaroo rats and white-footed mice, chipmunks are not found in burned areas of chaparral.

Rabbits and Hares (Order Lagomorpha)

Rabbits and hares are members of the family Leporidae. Three members of this family are found in chaparral: the Brush Rabbit (*Sylvilagus bachmani*), the Desert Cottontail

(S. audubonii), and the Black-tailed Jackrabbit *(Lepus califor-nicus),* also known as the Black-tailed Hare.

The Brush Rabbit is a true creature of the chaparral and is rarely found elsewhere. It has a head and body length of 11 to 13 inches, with charcoal-tinged brownish fur, and relatively short legs and ears. The Brush Rabbit tends to be nocturnal, rarely venturing beyond the dense brush, so that it is difficult to see. It rests and breeds in small openings, called forms, made in tangles of vegetation beneath shrubs. Its somewhat larger cousin, the Desert Cottontail (fig. 2), is found where chaparral is patchy, especially around the edges between thick shrubbery and openings. This is the rabbit that people are most likely to see in chaparral, as it is much less retiring than the Brush Rabbit. Both species feed primarily by grazing on grasses and other herbaceous vegetation. Their persistent feeding is an important cause of the general lack of herbaceous growth beneath and between mature chaparral shrubs.

The Black-tailed Jackrabbit is primarily a creature of the desert and open dry shrub communities. It is found in chaparral in recent burns and at the margins between shrublands and other plant communities, where it may use its long, easy bounds between shrubs to escape predators. It is large compared to the other two species, with long hind limbs, a short dark tail, and very large dark ears. The big ears dissipate heat and therefore help to keep the animal cool even under the hottest conditions.

Deer and Bighorn Sheep (Order Artiodactyla)

Two large herbivorous animals are found in chaparral: deer (*Odocoileus hemionus,* family Cervidae) and Bighorn Sheep (*Ovis canadensis,* family Bovidae). Both belong to the order Artiodactyla, a group of two-toed herbivores that also includes pigs and cattle.

Deer are common in chaparral throughout the state (fig. 40). The two subspecies living in southern California and the Sierra Nevada are called Mule Deer *(O. h.* subsp. *fuliginata*

Figure 40. A Mule Deer standing near a small toyon shrub. The large ears, as shown, pivot independently to pick up sounds.

and *O. h.* subsp. *californica)*, due to their large ears, while the subspecies of the North Coast Ranges and northern California is called Black-tailed Deer *(O. h.* subsp. *columbiana)*, because of the dark color of the upper surface of its tail. In addition to a large number of plant-eating rodents, deer are also important herbivores in the chaparral. They frequent chaparral after fire but are rarely found when the shrubs are fully developed and dense. This is because deer feed on grasses and forbs that are not present in the understory of mature chaparral, but that are a major feature of the postfire chaparral landscape in the first few years. They are seen even in mature chaparral; however, more forage is available along the edges of chaparral stands and in canyons. The density of deer populations in chaparral varies in time and space, with generally higher numbers in the northern part of the state. Deer are attracted to recently burned areas of chaparral, particularly when fires have been patchy, leaving open areas with lush vegetative growth adjacent to dense shrubs. As time passes, the capacity of burns to support deer gradually declines as the spreading shrubs close the gaps. If other chaparral patches are burned nearby, the deer herd as a whole can be indefinitely supported at a moderate level because a shifting matrix of edges between old and young vegetation is maintained.

When young growth is present, deer prefer certain shrubs, such as mountain mahogany *(Cercocarpus betuloides)*, ceanothus (*Ceanothus* spp.), and toyon, and generally avoid others, such as manzanitas (*Arctostaphylos* spp.). Under some circumstances, persistent browsing of preferred plants is sufficient to shape the structure and species composition of chaparral plant communities. In northern California and at higher elevations in the south, deer herds practice seasonal migration, moving in winter to lower elevations where there is less snow and more food. Some deer in southern chaparral move over shorter distances, following food availability by moving from south-facing slopes in winter and spring to north-facing slopes in summer and fall.

The Bighorn Sheep is a grayish brown animal with creamy white rump and horns (pl. 65). It is about the height of a smallish Mule Deer, but with a much chunkier body. The mature male has spectacularly coiled, massive horns that are used for mating combat with other males. These shy animals live in small groups, led by a dominant female, and they are likely to be seen only in remote, rugged areas. Once, two million Bighorns roamed across most of the mountain ranges of the West, but herds in California have dwindled to about 4,000 animals, found in scattered locations in the mountains of southern California, the eastern Sierra Nevada, and deserts. California's Bighorn Sheep are listed as an endangered species by the U.S. Fish and Wildlife Service, and as threatened by the California Department of Fish and Game. In the San Gabriel Mountains, Bighorn Sheep move down from the high mountains in winter and spring to the warmth and new growth of chaparral-covered canyons, where lambs are born.

In the late 1980s the California Department of Fish and Game moved 37 Bighorn Sheep from the San Gabriel Mountains to remote locations in the Sespe Wilderness of Ventura County, where Bighorns had disappeared almost a century earlier. After some initial difficulties, this transplanted herd seems to have increased in size. It occupies a large wilderness area with extensive chaparral. One reason for the small population size is probably the limited habitat value of the old and dense chaparral. Fire could improve the habitat for Bighorns by opening up the chaparral, by providing far more nutritious vegetation, and by improving the ability of the sheep to spot and elude approaching predators.

Plate 65. Bighorn Sheep. The male is the larger animal with curled horns, shown here with two females, photographed at the Living Desert, Palm Desert, California.

Carnivorous Mammals (Order Carnivora)

Carnivores, such as Coyotes, Gray Foxes, Bobcats, Mountain Lions, Black Bears, and Grizzly Bears are found in many communities across the country, and none is specific to chaparral. These animals often utilize the edges of chaparral where it is more open and where there are canyons and ravines associated with it. Despite the scientific classification as carnivores, not all of these species are strictly eaters of meat.

Coyote

One of the most successful and widespread mammals in North America is the Coyote (*Canis latrans*, family Canidae), and it is becoming more so with each passing year. A Coyote has the appearance of a medium-sized dog, resembling a small, unusually slender German shepherd, but with a more

Figure 41. A Coyote in recently burned chaparral, as suggested by the phacelia plant near the front feet. In colder climates it produces much thicker fur, especially in winter. Consequently, individuals living in chaparral tend to appear thinner and smaller than coyotes elsewhere in North America.

brownish back and rusty-colored ears and legs (fig. 41). The easiest method for detecting the presence of Coyotes is to listen for their cries, especially in the early hours of the evening. Coyote vocalizations are almost always produced by several animals at once and consist of a few brief howls followed by a series of short yips and barks. The howls of one group are sometimes answered by another group in the distance. These sounds are often mistaken for wolf howls, but there are no wild wolves in California. The Coyote is quite flexible in its diet and behavior, often including chaparral with other habitats in its search for food. In chaparral this animal is a hunter of the margins, ranging around edges and disturbed areas in search of rabbits, rodents, and other small animals. A Coyote cannot easily move through dense chaparral, so like us, it avoids it. Consequently, it is most often seen along roads, trails, railroad tracks, and when crossing openings. It is not uncommon to encounter Coyote scat made almost entirely of manzanita berries, which it eats for the semifleshy outer covering. However, the hard inner seeds pass through the digestive tract intact and are dispersed to new locations in the droppings. The tough manzanita seeds require heating by fire in order to germinate.

Some Coyotes, especially young males, will commute from chaparral several miles into cities and suburbs at night, either foraging or simply using fence lines, roads, and other thoroughfares to get from one place to another. These commuting predators have a reputation for favoring small pets and refuse for food, but our research has shown most of the diet of suburban Coyotes is wildland food.

Gray Fox

Found nearly everywhere in North America, the Gray Fox (*Urocyon cinereoargenteus,* family Canidae) within California is probably most common in chaparral (fig. 42). It frequents moderate to low elevations in heavy chaparral, canyons, and rocky places. It is the size of a small dog, with a very bushy tail,

Figure 42. A Gray Fox, one of the more common carnivores of chaparral. The very bushy tail is the most distinctive feature of this small canid.

about three feet from nose to tail tip. The Gray Fox is not frequently seen, probably because it is mostly nocturnal. As much scavenger as carnivore in its diet, this lithe animal moves easily through dense chaparral and will climb into shrubs to raid bird nests and take sleeping birds. Its droppings, which are seen much more often in chaparral than the animal itself, usually contain far more plant than animal material.

Mountain Lion

Also called Puma or Cougar, the Mountain Lion (*Puma concolor,* family Felidae) is a slender, tawny cat with a long, thick, black-tipped tail. Because of its large size, six to 7.5 feet, and distinctive tail, it is not easily mistaken for anything else, although a surprising number of startled people confuse it with the Bobcat *(Lynx rufus),* which is very much smaller and has a very stubby tail. It has the largest range of any mammal in the Western Hemisphere, from Canadian forests to southern

Chile. In California it occurs everywhere except the Central Valley and much of the deserts. It is most common where its principal prey, Mule Deer, are found. It favors chaparral, forests, rocky ledges and slopes, and mountain ridges. The Mountain Lion hunts by stalking and pouncing from ambush, so rugged topography gives it some advantage in leaping onto prey from concealed vantage points. This reclusive, solitary animal is rarely seen, even though it often lives near urban areas, sometimes even within city limits. This species' numbers have been steadily increasing in chaparral and elsewhere in California since hunting them for sport was discontinued in 1972, and the Mountain Lion is now occasionally seen on the edges of chaparral near urban areas.

Bobcat

Although sometimes confused with the Mountain Lion, the Bobcat (*Lynx rufus,* family Felidae) is much smaller, about the size of a very large house cat. A Bobcat has a stubby tail, broad cheeks, tufted ears, and stripes about the face, so that it resembles a domestic tabby (pl. 66). It occurs across most of the United States and Mexico, but in California, chaparral in

Plate 66. The Bobcat's short tail, facial marking, and ear coloring are distinctive, shown here photographed at the Arizona-Sonora Desert Museum, Tucson, Arizona.

Figure 43. A Bobcat in a typical rocky habitat.

steep and rocky terrain is a favored habitat (fig. 43). It is not often seen, even though it is frequently active in daylight. It spends most of its time lying in wait for passing prey, taking advantage of shrubs, rocks, and its camouflage coloration for concealment. A Bobcat's diet consists mostly of small mammals and birds, although given the opportunity it will readily take smaller and larger animals, including small domestic livestock.

Black Bear and Grizzly Bear

The largest mammal inhabiting California chaparral today is the Black Bear (*Ursus americanus,* family Ursidae), and it is also the one that occurs in lowest numbers. It occurs naturally throughout the chaparral and forests of the Sierra Nevada and the mountains of northern California, and widely throughout

North America. A Black Bear generally weighs 250 to 400 pounds, and despite the name, in much of its range the brown color phase predominates, varying from dark, chocolate brown to reddish brown to tan. Since the extirpation of the Grizzly Bear (*U. arctos,* family Ursidae) from California a century ago, the Black Bear has extended its range across the Tehachapi Mountains and northward into the Central Coast Ranges. It also occurs in the mountains of southern California, where it was introduced by the California Department of Fish and Game in 1932. In chaparral areas the Black Bear is found along the edges of dense brush and in canyons, where it forages for fruits and berries, insects, carrion, and whatever else might come its way.

At the time of the Gold Rush there was a second species of bear in California, the Grizzly Bear. It roamed the valleys and foothills of all parts of the state and was quite common in chaparral. Much larger than the Black Bear, a mature male probably weighed an average of 900 pounds. The last Grizzly in California was killed in the 1920s. This large, powerful, and unpredictable animal has always excited the imagination of Californians. The official State Prehistoric Artifact is a small carving of a walking bear, with the proportions of a Grizzly, made 7,000 to 8,000 years ago. During the Bear Flag Revolt in Sonoma, an 1846 insurrection against Mexican rule led by American settlers, the Grizzly Bear — as a symbol of great strength — was placed on a homemade flag. Soon thereafter, when California became a part of the United States, the Grizzly Bear was once again placed at the center of the state flag and in the foreground of the Great Seal of the state of California. The Grizzly Bear was chosen as a symbol of the independence, strength, and uniqueness of the state. Less than 75 years later it was gone from the California wilderness forever.

Inevitable as the extirpation of the Grizzly Bear might have been, its loss may have altered ecological patterns and processes of chaparral. Because of its numbers and size, this animal may have affected the very structure of chaparral by consuming fruits and distributing seeds, tearing up the soil to get roots and bulbs, and bulldozing openings in otherwise dense shrublands, and perhaps in other ways that are difficult to imagine now that they are gone. The Black Bear may have filled the ecological role vacated by the Grizzly to some degree. However, it is smaller and does not tunnel through the chaparral in the bulldozer fashion of the Grizzly Bear, so its impacts are clearly not the same. The presence of tunnels suggests that Grizzlies may have regularly opened up areas of chaparral that today remain closed and undisturbed except by fire. Large-fruited species of shrubs such as scrub oaks, cherries (*Prunus* spp.), and California coffeeberry are eaten by bears, and they do not require fire to germinate. These plants and others like them, may well have been dispersed and established intermittently in the middle of old chaparral by Grizzlies, a phenomenon that does not occur today. Contemporary footpaths cut through chaparral by humans often stimulate the growth of herbaceous understory plants such as soap plant *(Chlorogalum pomeridianum),* wild cucumber (*Marah* spp.), and virgin's bowers (*Clematis* spp.). Grizzly paths might have done the same. Moreover these paths undoubtedly increased access to dense chaparral for Coyotes and Mule Deer, and the openings would have provided suitable microhabitat at ground level for harvester ants (*Pogonomyrmex* spp.) and other invertebrates that require sun-warmed soil. The matrix of Grizzly Bear trails could have created and maintained a system for access, distribution, and colonization for a wide variety of plants and animals that are now excluded from extensive, continuous stands of mature chaparral.

The Grizzly Bear was also apparently attracted to areas of recently burned chaparral because these are places with an unusually rich and accessible supply of edible fruits, seeds, bulbs, and tender vegetation. Nineteenth-century visitors to California described remarkable scenes of as many as 20 Grizzlies peacefully congregated around a rich source of food. It is easy to imagine that patches of recently burned chaparral were the scenes of such bear conventions.

Birds

Approximately 50 species of birds are permanent residents of chaparral, and there is a similar number of seasonal inhabitants. The seasonal residents include species that come north from the tropics to breed in spring and summer, such as the Olive-sided Flycatcher *(Contopus cooperi)* and the Violet-green Swallow *(Tachycineta thalassina)*. Other species spend their winters in the chaparral and move to the mountains and farther north in summer for breeding. Several other species of birds, including the Western Scrub-Jay, Red-tailed Hawk, and Anna's Hummingbird are also common in and around the chaparral, but they also live elsewhere. The ubiquitous Bewick's Wren is also frequent in disturbed areas and along the edges of chaparral.

A small number of bird species make their home almost entirely in the chaparral. The two that are definitively chaparral birds are the Wrentit and the California Thrasher. The California Towhee and its more colorful relative the Spotted Towhee are consistently found where chaparral is mixed with other types of vegetation. The California Quail feeds around the edges and gaps between chaparral shrubs.

The diversity of birds in chaparral is due to the year-round availability of food. The most common food items for the birds of chaparral are insects and other invertebrates, and seeds and fruits. Insects peak in abundance during spring and summer, but some are present in all seasons. Most fruit and seed production also occurs in spring and summer, but other seasons have enough plant reproduction to support many seed- and fruit-eating birds as well. Trees and the herbaceous plants of chaparral often flower and fruit at times when shrubs do not, and some seeds are present on the ground and in the soil year-round.

Many of the bird species that inhabit chaparral shrubs are short winged, long tailed, and have moderately long legs and toes. These traits suit birds that fly only short distances, flitting through and beneath shrubs, landing on small twigs, and digging in the ground. To the dismay of inexperienced bird-watchers, as a group, most chaparral birds are rather drably colored in grays and browns. Bird-watchers sometimes refer to animals like this as "LBBs" (little brown birds). With a few notable exceptions, it takes more patience to locate birds in chaparral than in more open plant communities, and the lack of showy color patterns requires more careful observations to tell one species from another by sight.

This book describes those bird species most common and likely to be seen in the California chaparral. You are urged to consult one of the widely available field guides on birds for additional information.

Perching Birds (Order Passeriformes)

Wrentit

The most common and characteristic bird of chaparral is the Wrentit (*Chamaea fasciata,* family Timaliidae). The Wrentit is found in all dense chaparral, from southwestern Oregon to Baja California, and in adjacent plant communities with thick, shrubby vegetation. It feeds by flitting between branches, picking insects and fruits. This bird is rather plain in appearance, about six inches long, and gray brown. Its most conspicuous features are the upward cock of the slender tail and a light yellow eye (fig. 44). Wrentits are always found in pairs because they mate for life and permanently occupy a territory of .75 to 2.5 acres. They build cup-shaped nests in the midst of dense shrubs, well concealed from potential predators. The Wrentit is a weak flyer, but it can hop through the shrubs as rapidly as another species might fly. Since it is willing to fly only short distances, and reluctant to cross large

Figure 44. The tail cocked upward and the short wings are distinctive features of the Wrentit.

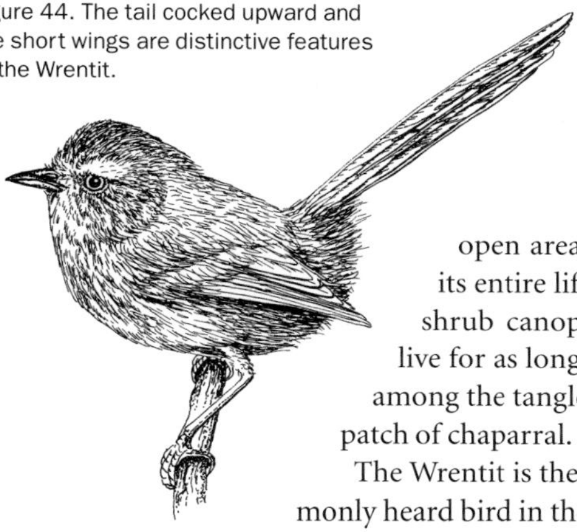

open areas, it spends its entire life under the shrub canopy and may live for as long as 10 years among the tangle of a single patch of chaparral.

The Wrentit is the most commonly heard bird in the chaparral. The male's call of three or four accelerating chips followed by a trill, all on the same pitch, is the background music of chaparral. It has been compared to the sound of a bouncing ping-pong ball. The female utters the chips without the trill. During spring, as many as a dozen singing males may be heard from one listening point, each busily proclaiming its breeding territory. A Wrentit will sing any time of the day and is often the only bird to be heard in chaparral outside of the breeding season. Heard a hundred times before it is seen, the surest way to spot a Wrentit is to sit or stand quietly in chaparral while producing some imaginative bird sound. A curious Wrentit will investigate the noise by hopping to a branch almost within reach, cocking its head from side to side, to inspect the noise maker with first one yellow eye and then the other.

California Thrasher

Like the Wrentit, the foot long, red brown California Thrasher (*Toxostoma redivivum,* family Mimidae), with its deeply downcurved bill and upraised tail, is a hallmark of chaparral (fig. 45). It inhabits all chaparral west of the deserts and below 5,000 feet, from the head of the Sacramento Valley through northern Baja California. It feeds on insects and seeds found

Figure 45. A California Thrasher feeding beneath California coffeeberry. The curved beak, and long tail and toes are useful field marks for identifying this chaparral bird.

in the leaf litter it stirs up with its bill. The Thrasher is a powerful bird that can move rocks and large objects in pursuit of its food. It has a distinct pattern to its foraging activities. First the bird uses its bill like a tiny miner's pick, striking the ground with a rapid series of excavating stabs, and then it switches to a side-to-side motion to clear away any loose dirt. This combination of movements exposes seeds and insects in the litter. The characteristically upraised tail helps to balance the bird during the pulling and digging of its feeding activities. When a desirable food item is located, the Thrasher may continue working away at it until the item is fully extricated. Thrashers eat seeds, fruits, leaves, and a variety of insects. Ants, wasps, and bees are particularly sought after, along with caterpillars, cocoons, and moths. The Thrasher eats many insects that humans consider to be destructive: tent caterpillars, wasps, beetles, and others that find their way into the Thrasher's stomach do not live to eat native or cultivated plants. A Thrasher will move out of chaparral to nearby gardens, where its insectivorous habits benefit the home gardener. This bird's bill grows continuously during its life and may become very long. It continues to grow to counteract the wear and tear of constantly digging in rocky soil. A young Thrasher has a relatively short bill that elongates noticeably after fledging. A Thrasher is a rather poor flyer, preferring to run along the ground or climb around in the bushes rather than fly. It rarely leaves the cover of the chaparral canopy except during courtship singing. You are much more likely to hear one throwing around leaf litter than to see one. California Thrashers have a protracted breeding season, from midwinter until midsummer. The nest, which is maintained by both parents, is well screened in the shrubs and only a few feet off the ground. The birds hold quite still while on the nest and follow circuitous routes to and from the nest, as defenses against predators.

Like its Mockingbird *(Mimus polyglottos)* relative, the California Thrasher is inclined to imitate the calls of other birds and may even use them in its defense. Thrashers have been heard to sing the songs of the Northern Flicker *(Colaptes auratus)*, House Finch *(Carpodacus mexicanus)*, California Quail *(Callipepla californica)*, Black-headed Grosbeak *(Pheuciticus melanocephalus)*, and Robin *(Turdus migratorius)* and can remember these songs for several months. They can also reproduce the howl of a Coyote *(Canis latrans)*, the call of a Red-tailed Hawk *(Buteo jamaicensis)*, the scream of a Western Scrub-Jay *(Aphelocoma californica)*, and the croak of a frog! Both males and females sing, and they prefer to do so near the tops of shrubs or small trees, which is the only time they are easily seen.

California Towhee and Spotted Towhee

The California Towhee (*Pipilo crissalis,* family Embarezidae) is frequently seen along the margins of mature chaparral and between shrubs in burned chaparral. It ranges throughout

California west of the deserts, around the edges of shrubby vegetation, including that found in suburban backyards. With a thick, conical beak and dusky brown coloration, this 8.5 to 10 inch long bird looks like an oversized, long-tailed sparrow (fig. 46). Paying little heed to nearby humans, and feeding around openings, it is one of the more easily seen birds of chaparral. The California Towhee digs in the litter for seeds, insects, and grubs, although not nearly as earnestly as the California Thrasher. It uses its feet together like a rake, scratching up seeds from around the shrubs with quick hopping motions. California Towhees build a bulky nest in shrubs, close to the ground. They seem to pair for life and spend their entire lives in a small patch of brush. The call of both sexes is a "chink," which is used to keep the pair together, and the song of the highly territorial male is the same sound repeated three or four times.

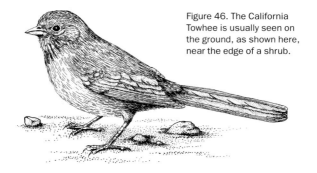

Figure 46. The California Towhee is usually seen on the ground, as shown here, near the edge of a shrub.

The Spotted Towhee *(Pipilo maculatus)* is a smaller, more colorful relative of the California Towhee. It has the same body shape as the California Towhee, but with chestnut colored sides, black head and throat, white belly, black back and wings with white spots, and red eyes. Despite its bright and distinctive coloration, it is seen less often than the California Towhee. The color pattern breaks up its silhouette when it is in dense chaparral and other vegetation. It forages in the thick leaf litter in the same two-footed fashion as the California Towhee. The Spotted Towhee is more inclined to perch on the outside of shrubs, and in small trees, than the California Towhee. You may see one when it moves up to the shrub tops to feed on seasonal fruits, or to sing during the breeding season. The nest is built by the female in dense shrubs, near or upon the ground. The call is a nasal "t-wee," and the song varies, often a trill in the cadence of "here-here-here-please."

Bushtit

The Bushtit (*Psaltriparus minimus,* family Aegithalidae) is a bird that feeds on insects found on chaparral shrubs. It ranges through all chaparral and other shrubby plant communities throughout the western United States. It is a very small bird, about four inches in length, with a longish tail and a short bill. Like most other birds that feed inside chaparral shrubs the Bushtit is plainly colored, with a gray back, a slightly darker cap, and pale underside. It gleans insects from the foliage by perching in almost any position, with the tail cocked this way and that, to best advantage for picking insects off the vegetation. Bushtits forage in loosely organized flocks of 15 to 60, constantly moving from place to place in little waves, while twittering to one another. Despite their small size and drab color they are relatively easy to spot, and even to count, because one tiny party of birds follows another in a regular pattern as the flock moves from shrub to shrub. Nests are finely woven pouches suspended from branches, seven to 12 inches long, made from grasses, leaves, and twigs, lined with spider webs and feathers. The call is a high-pitched twitter.

Bewick's Wren

The versatile and adaptable Bewick's Wren (*Thryomanes bewickii,* family Certhiidae) is found all over California and the other Pacific coast states, and across the southern United States, wherever there is dense shrubby vegetation with openings, such as chaparral and thickly landscaped backyards. This bird is about five inches long, gray brown on its back and paler underneath, with a white stripe above the eye and a long, thin, pointed bill that curves slightly downward. It frequently cocks its barred tail upward, and it switches it from side to side, rather than cocking it straight up like other wrens. In chaparral it forages within bushes, feeding almost exclusively on insects. Nests are found in a variety of places, usually cavities, and the highly complex song seems to vary between individuals.

California Quail

The California Quail (*Callipepla californica,* family Odontophoridae) is an easily recognized, plump, gray and brown bird about 11 inches long with a black, teardrop-shaped plume on top of its head. The male is boldly patterned in black, white, and earth tones with scallops and streaks. This bird is found throughout California and all along the Pacific coast from Canada to Mexico. The California Quail requires broken chaparral, with substantial and well-distributed openings between shrubs. It feeds in a style similar to chickens, pecking here and there on the open ground. It eats small seeds and tender plants in the openings between shrubs and uses the shrubs for cover and for roosting at night. The distribution of California Quail in chaparral is very uneven. Where there are sufficient openings, and where drinking water is available in the warm months, they are likely to be present. Where either of these resources is lacking, they are unlikely to be present. California Quail are always found in groups, called coveys, varying in size from one or two families of 10 to 20 individuals to as many as 200 birds. They much prefer to walk and run between shrubs when startled, but sometimes when a human or a predator draws near they will all take sudden flight, erupting with a startling whir of short wings, and then quickly dropping back down to safety a short distance away. They spend their entire lives within a small area, and they are unwilling to cross large openings to reach other patches of chaparral. Nests are built by monogamous pairs in small, lined depressions in the ground, concealed by vegetation, rocks, or other low-lying objects. While feeding they make low clucking sounds to stay together, so it is easy to tell when they are nearby, even though they are unseen. Their three-note call has the cadence of "ka-ka´-go," with the accent and a higher pitch on the middle note.

Western Scrub-Jay and Steller's Jay

Well named for its habitat, the Western Scrub-Jay (*Aphelocoma californica,* family Corvidae) inhabits dense vegetation wherever it is found. This includes the chaparral as well as other habitats such as riverside thickets and suburban backyards. This bird is noticed by everyone because it is noisy, relatively bold, and gregarious. It is just under a foot long, colored blue and blue black on the head and back, with ashy gray underparts and a thick, pointed bill (fig. 47). It is found in California, Oregon, and interior parts of the West in places with shrubby habitat, especially oaks. Omnivorous, it eats seeds, insects, carrion, lizards, small snakes and mammals, eggs and nestlings of other birds, and the odd French fry or other meal leftover. This is one of the few birds that can consume caterpillars covered with toxic, defensive hairs. It rubs the caterpillar in the sand, giving it a "haircut" before eating it. Western Scrub-Jays have been observed perching on the backs of cattle to pick off ticks and other parasites. Often miscalled a "blue jay" because it is blue and a jay, the Western Scrub-Jay is, however, a strictly western bird while the true Blue Jay *(Cyanocitta cristata)* occurs only east of the Rocky Mountains. At the upper margins of chaparral, where conifers appear, Western Scrub-Jays are replaced by the equally noisy and conspicuous Steller's Jay (*Cyanocitta stelleri,* family Corvidae), a much bluer bird with a black crest. In places where pines and chaparral are intermixed, the two jay species can occur within earshot of one another. The Western

Figure 47. A Western Scrub-Jay on a toyon branch. It is the only bird of its size in chaparral that is conspicuously blue on the outer wing, cap, and tail, shown here as dark gray.

Scrub-Jay call is a hoarse, single-syllabled, drawn-out rasp that rises toward the end, like a question, often repeated and echoed by other members of the flock. Nests are built in dense foliage, often near water.

Western Scrub-Jays live in small flocks of genetically related birds. They have highly organized social behavior that is most obviously manifested in their constant calling to one another. They will raid the acorn caches of other species of birds and take over backyard bird feeders by driving other birds away. They will collectively scold any intruder, human or otherwise, who is perceived to be threatening the welfare of any member of the group. Western Scrub-Jays, like their cousins, crows and magpies, are inquisitive and have intelligence reflected in quick learning and flexible behavior.

An example of their inquisitive behavior and intelligence comes from first-hand experience. In my (R.Q.) chaparral study site I set live traps for rodents, baited with seeds and peanut butter and arrayed in a regular checkerboard pattern. A flock of Western Scrub-Jays discovered that I was placing food in these traps in the late afternoon and quickly learned to follow along behind me, robbing each trap of its bait, and in the process springing the traps so that they were useless for my purpose. I tried changing my routine, setting traps at various times of the day, skipping several days, and setting them in different sequences, but to no avail. The birds were was always right behind me, calling to one another and, it seemed at the time, taunting me. Eventually, I had to resort to setting traps in the deepening dusk, with a headlamp, after the noisy and clever jays had retired for the night. One day one of these raiders was accidentally caught by one leg in one of the traps, uninjured but indignant. As I gingerly released it, trying to avoid injury to its leg, and to my hands from its vigorous pecking, several members of its flock perched just above my head, vociferously squawking about my treatment of their flock mate. If birds know how to curse, that is surely what they were doing.

Anna's Hummingbird

Anna's Hummingbird (*Calypte anna,* family Trochilidae) is the most common species of hummingbird found in chaparral, although several other species may be seen occasionally, especially during spring and late summer migration. It is found all year in the chaparral of both northern and southern California. Its original range is thought to have been the Pacific Slope of California from the Bay Area to Baja California, but that range has been extended northward to Canada and into southern Arizona and New Mexico. The adult male can be recognized by the iridescent magenta that covers the entire head. Like all the California hummingbirds it is quite small, three to four inches long, with a long, needlelike bill for probing flowers. In addition to feeding on nectar, Anna's hummingbird consumes more insects than the other North

American hummingbirds, taking insects from the air like a fly-catcher and also picking them off foliage while hovering. Anna's Hummingbird is among the first birds to breed each year in chaparral, beginning as early as late November and continuing through April. The tiny cup-shaped nests are lined with silk from spider's webs, and these nests may sometimes be found hidden in the dense branches of chaparral shrubs. The nearly year-round flowering of some native plant species, such as California fucshia *(Epilobium canum)*, and the early flowering of others, such as the golden currant *(Ribes aureum)* and fuchsia-flowered gooseberry *(Ribes speciosum)* (pl. 47) in canyons and at lower elevations, provide a natural source of nectar at all times of the year. It is hypothesized that the early breeding of Anna's Hummingbird in California developed in conjunction with these native plants. In and around urban areas these natural sources of food are strongly supplemented by ornamental plants and, of course, hummingbird feeders. This new, reliable, and widespread food is thought to be the cause of this species' recent range expansion. The male song is a loud and very distinctive high, rolling rasp with a mixture of equally high squeaks and little chirps.

Hawks (Order Falconiformes)

Hawks and owls are associated with the edges and openings in chaparral, especially where it borders woodlands and canyons that provide trees for roosting and nesting. Since these birds find food visually or by hearing, dense chaparral is not a place where they can easily hunt. They are more commonly seen flying above open, burned chaparral, where small mammal and large insect prey are common and relatively exposed, and along the edges between chaparral and more open plant communities. The most common hawks (family Accipitridae) associated with the chaparral are the Red-tailed *(Buteo jamaicensis)*, the Cooper's *(Accipiter cooperii)*, and the Sharp-shinned *(A. striatus)*. All three of these species have broad ranges that include most of North America.

Figure 48. A Red-tailed Hawk in flight. The dark belly band, in line with the center of the wings, is a useful field mark. The shading of the underside of the wing as seen in flight is variable, but the front of the wing is usually darker than the rest of the wing.

Red-tailed Hawk

The Red-tailed Hawk is often seen circling above chaparral looking for rodents, snakes, and rabbits in openings, and perched upon vantage points such as telephone poles and trees in search of rabbits, wood rats, pocket gophers, and other potential prey. When observed flying high overhead, the hawk is generally traveling, not hunting. This stout-bodied raptor is recognized in the air by its broad, rounded wings, four feet across, usually with a dark bar or wedge on the underside of the leading edge (fig. 48). The body plumage of the underparts varies from dark to light streaky brown, but the head is always dark. Despite this bird's name, the tail is not always red. An immature bird has a banded gray tail, and the red on the tail of a flying adult is only easily seen when the hawk banks and the upper surface of the tail becomes visible. Soaring hawks like the Red-tailed find their prey with keen vision. One of the major items of the Red-tailed Hawk diet in chaparral is the wood rat (*Neotoma* spp.), which although typically nocturnal, sometimes emerges from the nest at dusk. A Red-tailed Hawk that captures a wood rat must concentrate its attention on these abundant but elusive rodents during the passing interval of dusk, perhaps snatching one off the top of its nest or from an exposed branch. Red-tails have an unusually long breeding season, beginning in late winter and extending into summer. They build large, conspicuous nests on the tops of trees, telephone poles, and other places with unobstructed access from the top and a clear view in all directions. The nests, which may be as much as two feet across, are made mostly from large sticks. The call is a loud, descending "keeeeeer" with a hissing undertone that can be heard over considerable distances.

Cooper's Hawk and Sharp-shinned Hawk

The Cooper's Hawk and its smaller cousin, the Sharp-shinned Hawk, are slimmer and smaller than the Red-tailed, with short, rounded wings and long, banded tails. This body design gives them the maneuverability and speed to pursue and capture small birds such as the Bushtit between the chaparral shrubs. They fly quickly through the bushes, snatching birds out of the air and off perches. They are widely distributed in California and the rest of the United States. Around chaparral they are most likely to be seen in woodlands, silently flying away from the observer, straight through the branches of trees, or gliding just over the tops of the shrubs while watching for a careless bird. The Cooper's Hawk is about the size of a crow, with a slender body 14 to 20 inches long and a wingspan of 27 to 29 inches. It has a long, banded tail, rounded at its tip. The Sharp-shinned Hawk is the smaller of the two, with a body 10 to 14 inches long and wings that are 20 to 28 inches across. The tail is proportionally shorter than that of a Cooper's Hawk and is squared off or notched at the tip. Both species have gray brown backs and a lighter breast mottled with brown or red, a color pattern that allows them to blend in well with the light- and dark-speckled environment beneath trees and between shrubs. Although both species prefer small birds, and can empty a bird feeder in seconds as finches and towhees scatter in fear, they also take small mammals, reptiles, and even large insects. Birds are plucked, and mammals skinned, before they are consumed. They conceal nests in trees, which are defended fiercely even from a human passing nearby. The Sharp-shinned calls with a rapid series of high rasping squeaks, while the Cooper's call is similar but lower pitched.

Owls (Order Strigiformes)

The Great-horned Owl *(Bubo virginianus)*, Western Screech Owl *(Otus kennecottii)*, and some smaller species of owls with broad distributions are also found peripherally associated with chaparral. Since these owls require openings between shrubs to hunt, they are uncommon in mature chaparral. Like the hawks, the first few years after fire when rodents are plentiful and relatively exposed, owls will be more abundant. The surest way to know whether owls are present, and which ones, is to listen at night for their distinctive calls. This is best done in late winter to early spring when they are about to breed and are calling to proclaim breeding territories. Calls can be heard over considerable distances, and in chaparral might come from nest sites or high perches in nearby woodlands.

Reptiles

Snakes and lizards are important to the overall well-being of chaparral communities statewide. These animals keep the populations of seed-eating rodents in check and also keep insect populations under control. They are an important food item for some predators, and indeed some snakes and lizards eat other reptiles. These complex interactions help to balance numbers of plants, herbivores, and carnivores in chaparral, thus assuring stability and continuity of the community. There may be as many as a dozen species of reptiles in a given area of chaparral, more than in the majority of other ecological communities of the state. These snakes and lizards typically have wide habitat preferences. Rattlesnakes, Gopher Snake, Western Fence Lizard, Side-blotched Lizard and occasional alligator lizards are among these widespread species. Two reptiles closely adapted to chaparral are the California Whipsnake and the Coast Horned Lizard. Most species are seldom seen because they are nocturnal and secretive, or are inactive when humans are most likely to visit. In addition to the snake species described below, about a dozen additional species of snakes are known to occur in chaparral. These are rarely seen, either because they are actually scarce in chaparral or because their habits and habitats make it unlikely that anyone, other than a determined herpetologist, would find them. In chaparral most of these elusive species occur only in southern California.

Snakes (Order Squamata, Suborder Serpentes)

Western Rattlesnake and Red Diamond Rattlesnake

For many people exploring chaparral, the first reptile that comes to mind is the rattlesnake. The Western Rattlesnake (*Crotalus viridis,* family Viperidae) is a ubiquitous inhabitant of chaparral and most other natural environments of California west of the deserts. The surest way to recognize this animal is the broad triangular head that is wider at its base than at the neck and of course, the rattles at the tip of the tail (fig. 49). This snake can grow to lengths of as much as five feet, although 1.5 to three feet is more common, with the young barely a half foot long. Contrary to popular legend, the number of rattles does not correspond to the age in years of any

Figure 49. A Western Rattlesnake shown coiled in resting position. If the snake were alarmed the head would be raised higher and pulled back over the center of the coil, and the tail rattle would be turned up vertically.

rattlesnake. Additional rattles are added whenever the snake sheds it skin. Shedding occurs as the snake grows, and that can happen several times in a year. The color of the Western Rattlesnake varies considerably from place to place, with mottles of brown, greenish colors, and black superimposed on an overall color that often is grayish. The Western Rattlesnake is active during the day or night, depending on the temperature, hunting small rodents such as kangaroo rats (*Dipodomys* spp.) and white-footed mice (*Peromyscus* spp.). This snake patrols a home area, following animal scent trails and curling up and waiting in places where prey is likely to pass by. It detects the location of potential prey from body heat, which it accomplishes by means of special sensory organs located in pits between the nostril and eye. The heat pattern allows it to accurately strike at prey, injecting venom that quickly paralyzes and kills the animal. The Western Rattlesnake is most active from April to October and hibernates during most of the rest of the year.

The beautiful Red Diamond Rattlesnake *(Crotalus exsul)* inhabits chaparral and other brushy habitats in the Peninsular Ranges of southern California from Orange and Riverside Counties southward. It has the same general body size and shape as the Western Rattlesnake, but as the name implies, it has a diamond pattern on the back and is reddish or tan in coloration overall, with pronounced dark and light rings on the tail.

The most important thing to remember about rattlesnakes encountered in chaparral is that they have a retiring disposition and, when given a choice, will avoid or move away from contact with humans. If a person or any other large animal draws too near, a rattlesnake will begin to rattle by vibrating its tail. This sound is a warning, designed to ward off an approaching animal. Heed this warning! If you hear a rattling snake, freeze immediately until you determine the location of the sound, and then slowly move away from the source. It is best, when moving through chaparral or any other dense vegetation, to watch where you step and to move slowly enough to give a startled rattlesnake time to raise a warning before you draw too near. In chaparral there is little cause for concern in the cooler months, when rattlesnakes are hibernating.

California Whipsnake

The California Whipsnake (*Masticophis lateralis,* family Colubridae), also known as the Striped Racer, has been called the "chaparral snake" because it is so characteristic of this habitat. Its range almost exactly coincides with that of chaparral. It frequents the rocks and bushes of the chaparral, eating a variety of small mammals, lizards, other snakes, birds, and insects. It is a slender snake that grows up to five feet long, brownish or black on its back, with a yellow or white stripe on each side (fig. 50). The California Whipsnake is well adapted to the chaparral habitat. In addition to foraging on the ground,

Figure 50. A California Whipsnake in hunting posture. Note the slender body and head, and the light stripe along the flanks.

it also spends much of its time in the shrubs basking on the upper branches, hunting for eggs and other prey among the stems, and seeking shelter in the larger branches. While hunting it holds its streamlined head above its body. The constant head motion and prominent eyes give this reptile a very alert appearance. It is efficient at exploring burrows and other hiding places of its prey. As shown in pl. 67, it seizes live prey, which it holds in its powerful jaws until it stops struggling, and then swallows.

Temperature regulation is exceptionally well developed in the California Whipsnake, which is one reason it is so successful in chaparral. It is active from March to October and then hibernates until spring. This snake is often noticed because it is active during the heat of the day and moves about quite quickly beneath chaparral shrubs, as the name "racer" implies. It is too nimble to be easily captured and can deliver a harmless bite—so quickly that the pursuer may be unaware for a time that it has been bitten. Sometimes this snake emits a musky smell when handled.

Plate 67. A California Whipsnake swallowing a Western Fence Lizard. The snake temporarily unhinges its jaw so that the lizard can pass through its throat into the digestive tract. This leaves a food bulge in the snake's body that gradually shrinks and moves toward the rear as the lizard is digested.

California Whipsnakes mate in late spring and lay eggs about a month later. The eggs hatch in October or November. Small snakes heat and cool more rapidly than their larger parents, so the hatchlings can continue to be active on warm days later into fall than can the adult snakes. They depend on getting a good store of food into them before winter cold stops their feeding. Perhaps it is this active period early in their lives that allows them to grow particularly rapidly. One-year-old snakes are 12 to 18 inches long, and that length is doubled by the time a snake is two years old. After two years, growth is slower. It is difficult to determine the ages of adult snakes, but they are thought to live to about five or six years.

Gopher Snake

The Gopher Snake (*Pituophis catenifer,* family Colubridae) is a large, light to medium brown, heavy-bodied snake common in many habitats across the western United States. It occurs occasionally in the chaparral, where it frequents openings between shrubs. It can grow up to six feet long, although it is ordinarily about half that size, and is a yellowish or creamy color with blotches of brown, black, or reddish brown on the back. The large size and daytime habits make this an easily noticed animal. When a Gopher Snake is disturbed it will flatten its body, hiss, and vibrate its tail. Despite these theatrics, it is harmless and relatively docile. A Gopher Snake will come out onto paths and roads for warmth. In this conspicuous location, the diamondlike pattern of blotches on the back and its habit of tail-shaking when disturbed sometimes cause it to be mistaken for a rattlesnake. This reptile is beneficial in its natural habitat and around homes because it controls rodent pests, as is implied by its name, and so it is best left alone. In spite of its large size, a Gopher Snake can climb, but its preferred method of hunting is to crawl down burrows and corner rodents. It kills larger prey by constriction but may simply swallow smaller animals. It hibernates below ground in the colder months, breeds in spring, and lays eggs in early summer, which hatch later that same season. Newly hatched Gopher Snakes are about a foot long.

Western Patch-nosed Snake

The Western Patch-nosed Snake (*Salvadora hexalepis,* family Colubridae) is widespread in the deserts and chaparral of southern California but is not often encountered. Typically about three feet long, it has a slender body with dark sides and a broad yellow or tan stripe down the back. It derives its name from an enlarged scale that folds up and back across the tip of the snout, just above the mouth. Presumably this scale assists the animal in burrowing and pushing under rocks. Like the California Whipsnake it moves quickly, is day-active, and preys on lizards, buried eggs, and small mammals in their burrows. It will also climb in search of bird nests. It remains active whenever there are warm days. Eggs are laid in early summer, with hatchlings just under a foot long emerging in late summer.

Lizards (Order Squamata, Suborder Lacertilia)

As with snakes, few species of lizards are restricted to the chaparral. Most are found in a variety of other open and brushy habitats across the state. The most nearly chaparral-specific are the Coast Horned Lizard and the Western Whiptail. Other common chaparral lizards in the family Phrynosomatidae are the ubiquitous Western Fence Lizard, the Sagebrush Lizard, and the Side-blotched Lizard, as well as the Southern Alligator Lizard (family Anguidae).

Coast Horned Lizard

The Coast Horned Lizard (*Phrynosoma coronatum,* family Phrynosomatidae), also sometimes called Horny Toad, is characteristic of the chaparral and adjacent communities from northern California through Baja California. It prefers sandy soil in which it can dig down for shelter. Its dragonlike appearance is unmistakable, with a row of sharp horns at the back of the head, a round, flattened body, spiny scales along the back and tail, and two rows of smaller spines on each side of the body (fig. 51). The head and body are 2.5 to four inches long. This lizard feeds mainly on native harvester ants (*Pogonomyrmex* spp.), supplemented by other insects. The harvester ants in turn depend on the seeds of chaparral and other native plants for their food. These feeding relationships link the survival of the plants and animals. The Coast Horned Lizard is classified as a Species of Special Concern by the California Department of Fish and Game and the U.S. Fish and Wildlife Service. It has disappeared from most of the lower-elevation portions of its original range due to habitat loss, and quite possibly the disappearance of harvester ants, which have been exterminated in many places by the aggressive, exotic Argentine Ant *(Linepithema humile)*. The Coast Horned Lizard does not eat Argentine Ants. This loss of food could be further aggravated by the recent arrival in coastal California of another aggressive, imported species, the Fire Ant *(Solenopsis invicta)*. Given these losses and displacements, remaining habitat for the Coast Horned Lizard in chaparral would be in places away from urban and agricultural disturbance, the exotic Argentine Ant that often accompanies human activities, and marauding house cats.

Figure 51. The Coast Horned Lizard, increasingly rare in chaparral near populated areas, is the only horned lizard of chaparral.

The Coast Horned Lizard utilizes open areas between shrubs, which is where harvester ants are likely to be found, and sandy soils in which it can bury itself. In chaparral such openings are commonly near the bottoms of washes, along roads, and on uneven terrain. It spends the hottest parts of the day and much of the late afternoon and evening buried from one to several inches underground, where the soil is cooler and moister than on the surface. The Coast Horned Lizard hunts by situating itself near an ant nest or trail and waiting for ants to walk by. As ants parade past an effective tongue flip dispatches them quickly and quietly. This method is efficient, as it limits energy spent on hunting. A small lizard requires only about one-thirtieth of the food needed by a small bird of the same size. This lizard is most active between April and August, which is the time of year when harvester ants, its preferred food, are out and about.

Coast Horned Lizards lay eggs in late June. The young hatch in September and forage for a period of time in the early fall. Adults become largely inactive after egg laying, and remain in burrows or buried in the soil during the hottest part of summer. This inactive state during the summer heat gradually grades into winter hibernation, from which they emerge in early spring.

One of the most interesting adaptations of the Coast Horned Lizard to life in the open in a sunny climate is an internal sunscreen. To prevent damaging ultraviolet light from reaching the internal organs, which are just a few tenths of an inch below the skin, the lizard's peritoneum, the membrane lining the body cavity, is heavily pigmented. This membrane is almost black and thus screens out the light. This unusual color adaptation in the peritoneum is in marked contrast to most other animals, including humans, where the membrane is a bright pink. The darkly pigmented lining functions like internal "sunglasses."

Because Coast Horned Lizards spend much of their time in the open, they are a likely meal for more active predators such as snakes, other lizards, or birds of prey. If approached, the lizard will tend to freeze, relying on protective coloration to avoid detection. If a Coast Horned Lizard is lightly touched, it can scurry away to safety as quickly as any other lizard. If protective coloration or flight does not suffice, this lizard has other options. Some of these are mechanical, such as its spines, while others are behavioral such as rising up to look larger, but the most unusual involves shooting blood at its pursuer!

When a dog, coyote, fox, or persistent human closely approaches a Coast Horned Lizard, it may turn to a chemical defense. It can use blood vessels and sinuses in its head to increase blood pressure around the eyes, building it up until blood is forced out the tear duct from between the closed eyelids. The dog, fox, or coyote luckless enough to annoy the lizard at this moment is then forcibly sprayed with fine jets of blood, which they find quite repulsive. It is not known why doglike predators find the blood so distasteful, but the reactions of the unlucky canines make obvious that it is very unpleasant. The Coast Horned Lizard can perform this defensive blood squirting several times if hard pressed. This treatment usually deters the offending predator.

Western Whiptail

The Western Whiptail (*Cnemidophorus tigris,* family Teiidae) is a very common lizard in chaparral as well as in other habitats across the state. It is slender, with a pointed snout and a tail about twice as long as the 2.4 to five inch head and body (fig. 52). The back is reddish brown with dark blotches and as many as eight light stripes, sometimes indistinctly defined. It

Figure 52. A Western Whiptail, with a small California poppy plant in the background. Constant motion makes this lizard easy to detect, and the long tail makes it easy to identify.

has long hind legs and toes and drags its tail as it scurries from place to place. It is often heard before it is seen, scuttling through dry leaves. The Western Whiptail has a fidgety manner about it and can be recognized by its constant, jittery movements as it forages for invertebrates in the leaf litter. This lizard is active all day in chaparral even when the weather is quite hot, constantly twitching its head from side to side. It is the principal food item of the California Whipsnake (pl. 67), and the two are active throughout the day. Eggs are laid in late spring, and both adults and young enter hibernation in late fall to early winter.

Western Fence Lizard and Sagebrush Lizard

The Western Fence Lizard (*Sceloporus occidentalis,* family Phrynosomatidae) is one of the most numerous and widespread species of reptile in California, found everywhere except deserts and mountains above 6,000 feet. It has a brown, black, or light gray back, with blue along the sides of the belly, and is about six inches long including the tail (fig. 53). In spring the throat and belly patches of the male become a brilliant blue, and the male displays as much color as possible to other lizards by doing pushups while distending its throat. Viewed in profile, which is the perspective of another male lizard, this posturing emphasizes the size and blueness of the posturing male. This lizard's name is derived from its habit of using vertical wood, rock, or masonry surfaces for sunning, so that near human habitations you can see it often on fences, walls, and wooden posts. In mature chaparral, it is most often found around openings in the shrubs, and other places where the sun reaches the ground (pl. 31). It can be found out and about on any warm and sunny day, even in winter. The Western Fence Lizard forages for all sorts of invertebrates on the ground, rock surfaces, and branches. The high perches selected for foraging, and for territorial displays by males, make it easy prey for snakes and hawks. Breeding occurs in spring, and six to 15 eggs, one-quarter to one-half inch long, are laid

Figure 53. A Western Fence Lizard on a burned chaparral shrub stem. A very common lizard in sunny places within and around chaparral.

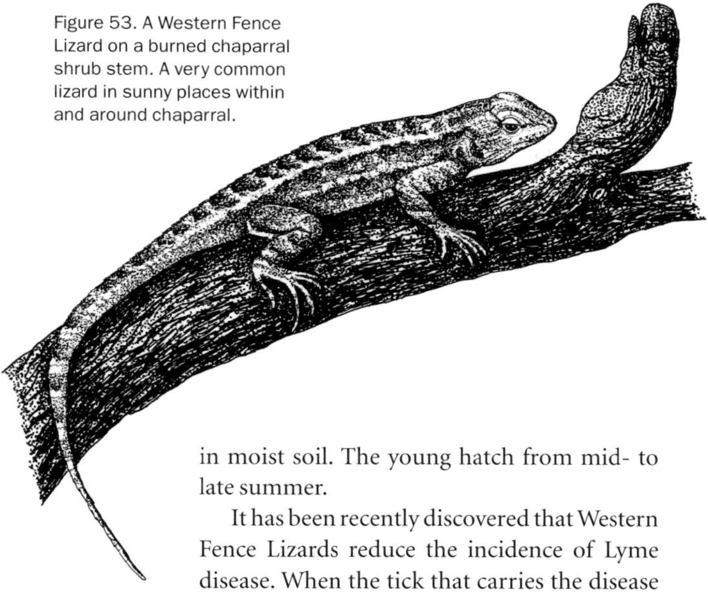

in moist soil. The young hatch from mid- to late summer.

It has been recently discovered that Western Fence Lizards reduce the incidence of Lyme disease. When the tick that carries the disease bites these lizards, apparently, the disease-causing bacterium within the ticks is killed. The lizards seem to be immune to the disease, and this immunity is transferred to ticks along with the lizard's blood. About five percent of ticks living in areas with Western Fence Lizards carry Lyme disease, while 50 percent carry the disease in places without these lizards. No one would have guessed that a lizard would have anything to do with this disease.

The Western Fence Lizard can survive chaparral wildfires by staying below ground, after which they are often seen sitting out on the ends of burned branches. The Sagebrush Lizard (*Sceloporus graciosus,* family Phrynosomatidae) is a similar but slightly smaller animal that occupies chaparral at elevations above 5,000 feet. This is primarily a mountain animal, ranging up into pine forests, but at intermediate elevations the ranges of Sagebrush and Western Fence Lizards overlap.

Side-blotched Lizard

The Side-blotched lizard (*Uta stansburiana,* family Phryno-somatidae) is a very common small lizard of open chaparral as well as many other habitats in central and southern California. The head and body are about 2.3 inches long, and the tail is somewhat longer. It has a dark blotch just behind the front leg on both sides, a mark that makes it relatively easy to identify and provides its common name. The male often has a lot of yellow to orange on the sides of the head and body, and blue speckling on the back. In chaparral it is most common in openings where the vegetation has recently burned or otherwise been disturbed, and on very steep, rocky slopes. It is absent from dense chaparral. Active all year in most places, it pursues a variety of invertebrates taken from the ground, rocks, and lower branches of shrubs. Breeding and egg laying take place from spring to late summer, and a female may produce several clutches in a single season.

Southern Alligator Lizard and Northern Alligator Lizard

The Southern Alligator Lizard (*Elgaria multicarinata,* family Anguidae) and Northern Alligator Lizard (*E. coeruleus*) are slender, long-bodied animals with a head and body up to six to seven inches long, and a tail that can be twice that long. They have a glossy look due to smooth scales. A large triangular head together with small legs gives them a snakelike appearance. These Alligator Lizards are reddish to gray brown, with dark stripes across the back. They are found everywhere in the state except the deserts and high mountains. The northern species is slightly smaller than the southern, but the two are difficult to tell apart without close inspection of the scales. This is not recommended, because when handled they will invariably turn and deliver a painful bite with their surprisingly powerful jaws. The range of the two species overlaps both along the central coast and in the foothills of the Central

Sierra Nevada. They prefer relatively dense cover, so that in and around chaparral they would be most often encountered in weedy vegetation, and in places with an accumulation of leafy litter, logs, or other objects under which they can hide. They eat insects, small rodents, birds, and even other lizards. They climb quite well, using their tail for grasping and balance, and sometimes nest in the crotches of trees that have accumulated leaves and twigs. Both are common residents around houses and frequently found in backyards, particularly if a pool of water is present. Breeding takes place in late spring, eggs are laid in midsummer, and eggs hatch about two months later, producing young no more than three inches long.

Amphibians

Chaparral is generally considered too dry for amphibians. Since they require moisture for reproduction, adjacent habitats such as oak woodlands and watered canyons contain more species and more individuals. The lack of moisture in chaparral is especially limiting in southern California, where damp environments are fewer and farther between than they are in the north. In chaparral the best time to search for amphibians is on temperate winter or spring nights during rain, especially in canyons and other places not too distant from water. During the drier part of the year amphibians retreat to cool, moist refugia such as wood rat nests or migrate to places that have damp soil and litter, such as watered canyons. The most likely amphibians to be encountered in chaparral are the Ensatina Salamander and the Western Toad.

Ensatina Salamander (Order Caudata)

The Ensatina Salamander (*Ensatina eschscholtzi*, family Plethodontidae) is sometimes found in chaparral, although it is more commonly associated with redwood forests and other coniferous forests of northern and central California. The head and body range from two to three inches in length, plus a thick tail that is constricted at its base (fig. 54). The color varies greatly from place to place but is generally dark with yellow, orange, or pink spots that may fuse together to various degrees. The smooth skin must remain relatively moist, because the Ensatina lacks lungs and must breathe through its skin. In chaparral it is only during the rainy periods that it can move freely about, foraging for various invertebrates and snails. Mesic and northerly preferences notwithstanding, this animal inhabits chaparral even in southern California. One reason this salamander is able to inhabit chaparral is that it does not reproduce in water, laying eggs in moist places or in surface litter during the rainy season. The young hatch as small adults. This trait is shared with a number of other

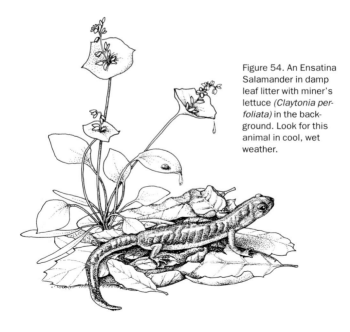

Figure 54. An Ensatina Salamander in damp leaf litter with miner's lettuce *(Claytonia perfoliata)* in the background. Look for this animal in cool, wet weather.

species of California salamanders, which sets them apart from other North American amphibians that lay eggs in water. Ensatinas take several years to reach sexual maturity and live as long as 15 years.

Western Toad (Order Anura)

The Western Toad (*Bufo boreas,* family Bufonidae) is found almost everywhere in California west of the deserts. The body is 2.5 to five inches long and colored grayish to greenish, with the warts set upon blotches that are dark and often rusty colored. A cream-colored stripe runs down the middle of the back, from between the eyes to the rump. The Western Toad tends to walk rather than hop. Unlike the Ensatina Salamander, this animal is bound to water for reproduction, so it can probably venture no farther into chaparral than the distance it can travel from the nearest body of surface water. At the extreme this commute may be about 1.2 miles in each direction. It doubtless retreats below ground into rodent burrows and other moist recesses during the long summer stretches of warm, dry weather.

Insects and Arachnids

Insects and their relatives (spiders, mites, ticks, and scorpions) show some of the most specialized adaptations within the chaparral community. Adaptations are traits that allow an organism to live successfully in its particular environment. These adaptations can be physical, physiological, or behavioral. Others are inextricably linked with plants for their reproduction and survival. Below are some interesting examples of adaptations to living in chaparral. Other insects are discussed in chapters 2 and 3, and earlier in this chapter under Wood Rats. Much more remains to be learned about these numerous and important organisms, so this is but a small selection of accounts about some better-known species.

Insects (Class Insecta)

Harvester Ants (Order Hymenoptera)

Ants are among the most obvious ground dwelling insects of the chaparral. A survey at four chaparral sites found 45 different kinds of ants, about one fifth of all the ants known in California. One of the best-known and conspicuous ants in the chaparral are the harvester ants (*Pogonomyrmex* spp., family Formicidae). These ants have conspicuous nests in clearings between shrubs in mature chaparral and upon the blackened soil in burned chaparral. Harvester ant nests typically have a cleared area up to a yard wide around the nest entrance, with a ring of chaff at the outer edge. The ants themselves can be up to one-quarter inch long and are bright red with conspicuous bristles on the legs and abdomen (fig. 55). While they are not extremely aggressive, disturbing a nest will produce a swarm of stinging, biting ants, and these bites and stings can be quite uncomfortable for several days. They have the habit of crawling up human legs and stinging wherever they get stuck, so it is unwise to stand still for long where harvester ants are busy.

Harvester ants eat seeds of chaparral plants. A foraging worker ant will travel 100 feet or more from the nest to find food. They search for seeds during the day using vision, so they are most active in the mornings and afternoons, frequently staying underground during the hottest part of the day. As described earlier in this chapter, the Coast Horned Lizard *(Phrynosoma coronatum),* their principal predator, also hunts at the same time. Harvester ants patrol the ground around fruiting shrubs and, in some cases, seem to pick up seeds the moment they fall upon the ground. Once a worker has a seed it returns to the nest site to shuck it. The ants leave the chaff outside the nest, forming a conspicuous ring of hulls, and take the cleaned seeds into the nest. As illustrated in fig. 55, harvester ants will grasp surprisingly large seeds, sometimes as big as the ant's head. They can do this because of the wide gape of their jaws and muscles that work the jaws like an exceptionally strong pair of tongs. When the ants bring seeds into their nests, they are not only provisioning themselves, but they are also incidentally protecting seeds from decay and from being eaten by other organisms. Some seeds may remain in the underground storage chambers until the next fire, and be in a very good place to germinate and grow when rain falls on the burned area.

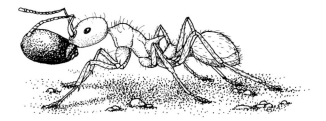

Figure 55. A harvester ant carrying a ceanothus seed by grasping it between its jaws.

Harvester ants are not active in winter, so the nest entrances usually remain closed during that time, but the nests are still readily identified by the circle of cleared ground with a ring of discarded material around the outside. Their nest sites require the warmth provided by direct sunlight, so they are excluded from places where chaparral shrubs shade the soil.

Harvester ants are particularly abundant after fire, in response to the greater number of seeds available from the fire annuals that flower so abundantly then, and the expanses of unshaded soil created by the fire. Some fire-following plants, such as the bush poppy *(Dendromecon rigida)* (see chapter 4), have a special relationship with harvester ants. The bush poppy produces seeds with structures called elaiosomes attached to them. The elaiosome is an oil- and nutrient-rich structure providing immediate energy for the ants. The ants carry the seeds back to the nest using the elaiosome as a handle, and they receive a food reward for this activity. The ants typically remove the elaiosome first and tend to the seed later. This is a mutually beneficial relationship between ant and plant since the elaiosome feeds the ant, which in return disperses bush poppy seeds away from their point of origin. This is a common arrangement between ants and plants in the shrublands of South Africa and Australia, but it seems to be unusual for chaparral. The large carpenter ants (*Camponotus* spp.) also move seeds about in chaparral, but these ants are found in mature chaparral, where they are more difficult to detect.

Antlions (Order Neuroptera)

The bodies of chaparral ants are a good source of water and nutrition, so they are the chief food items of the Coast Horned Lizard *(Phrynosoma coronatum)* and also of the larvae of a predatory insect called an antlion (*Myrmeleon* spp., family Myrmeleontidae). An antlion, often called a doodle bug, is actually the larva of a beautiful winged insect that resembles a small mayfly or damsel fly. The adult antlion's only function in life is to reproduce. It does not feed, but exists in a winged state for the short-lived courtship flight and mating. The feeding stage is the larva, which is a voracious predator. The antlion larva makes a trap with its paddle-shaped head, which is specially shaped for digging and flipping sand. It catches and dispatches its prey with a pair of powerful hooked jaws.

Antlions are found where the soil is fine textured and dry. The antlion makes a cone-shaped depression with the sides sloped at a precise angle such that tiny soil particles or sand grains will slide downward at the slightest disturbance. Once an ant or other insect steps on the edge of the pit, the disturbed grains begin to slide toward the bottom. With each step more soil slides down the sides, carrying the hapless insect ever deeper in the pit. The antlion, meanwhile, actually throws sand at the struggling victim, causing a yet greater cascade down the sides. At the bottom of the trap, the waiting antlion larva seizes its meal with dispatch. Once the larva is full grown it spins a cocoon underground, goes though pupation, and emerges as a pale, diaphanous, night-flying adult. By the end of its one evening of courtship, eggs are laid and a new generation of antlions begins its development.

Gall Wasps (Order Hymenoptera)

Members of the gall wasp family (Cynipidae) use plants to provide food and shelter for their larvae. Oaks in particular are a favored host for these wasps. They deposit their eggs in plant stems and leaves after piercing the plant surface. Chemicals associated with the eggs, and later with the larva, cause the affected part of the plant to gradually enlarge to form a structure called a gall, within which the insect grows. The exact mechanism by which these chemicals stimulate the plant to produce abnormal growth is not fully understood. On oaks (*Quercus* spp.), the galls also contain protective chemicals called tannins, which likely serve to defend the developing gall wasps as well as they defend the oak.

Of the galls found in the chaparral, oak "apples" are among the easiest to observe and are often seen on the stems of scrub oaks (pl. 68). As shown in pl. 68, they are sometimes quite red and larger than a golf ball and may exude a sticky sweet-

Plate 68. A wasp gall growing from the stem of a scrub oak.

tasting substance. The gall is filled with a spongy tissue that is rich with water and sugar from the plant, and this is what nourishes the larva within. If you cut a fresh gall open in summer you may find the small, cream-colored, wormlike larvae inside. In contrast, most galls on other chaparral plants are as insignificant in appearance as simple curls or bubbles on a leaf, or bumps on a stem.

Gall wasps have some peculiar traits. For instance, one generation may have both males and females with wings, a rather natural thing for a wasp, but the next generation will be nothing but wingless females that have been produced asexually, meaning that the eggs hatch without fertilization by a male. Another oddity is that different species of wasps may sometimes share the same galls. This happens because some gall wasps lay their eggs in galls that have already been formed by other species of wasps. In the end, two different species of wasps will emerge. Galls are fascinating to observe. As is the case with many aspects of the biology of chaparral insects, they need much more study before they will be well understood.

Plate 69. A Red-haired Velvet Ant. The bright coloration advertises its presence as it moves on the ground.

Velvet Ants (Order Hymenoptera)

"Velvet ant" is the common name applied to a group of wasps (family Mutillidae) in which the females lack wings. They attract attention because they are fuzzy and brightly colored and are often seen scurrying over open soil around chaparral in late summer. Several species occur commonly in chaparral, including the Red-haired Velvet Ant *(Dasymutilla coccineohirta),* shown in pl. 69, and the White Velvet Ant *(D. sackeni).* They are about one-half and three-quarters inch long, respectively. The biology of velvet ants is not well understood, but they probably parasitize the larvae of ground-nesting bees and wasps. When seized, velvet ants make a characteristic squeaking noise. Do not use your fingers to grab a velvet ant because they have a very painful sting. In other parts of the country they are called cow-killer ants, and in Mexico *hormigas del diablo* (devil ants).

The Yucca Moth (Order Lepidoptera) and the Chaparral Yucca

One of the most interesting relationships between plants and insects in the chaparral is seen between the Yucca Moth *(Tegeticula maculata,* family Prodoxidae) and the chaparral yucca *(Yucca whipplei).* This pair has evolved together over a long period of time, and the survival of each is secure only if they work together. Yuccas are a bit of a conundrum to botanists, because we cannot age them and we do not know what causes them to flower, or prevents them from flowering. We observe, however, that when one chaparral yucca blooms, so usually do many others in an area, and this is particularly so after fire (pl. 9). A fire will sometimes stimulate yuccas to flower months out of the normal season of June. Yuccas and Yucca Moths are widespread in the western United States, but the chaparral yucca has its very own moth, found nowhere else, and that moth can survive only by rearing its young in this chaparral yucca plant. The yucca plant produces pollen that must be transferred from one flower to the next so that fertilization can occur and seeds produced. This is where the Yucca Moth comes in, because only the Yucca Moth can do the job for the yucca plant.

Normally, flower pollen is small and easily dislodged on

insects as they fly or climb about the flower. Yucca pollen, however, is sticky and hard to remove from the anther, the male flower part that produces it. Yucca pollen comes loose as a series of sticky strands. From most insects' point of view, chaparral yucca pollen has all the properties of chewing gum on a hot sidewalk, a substance to be avoided. The female Yucca Moth, however, has just the right equipment to handle the sticky pollen. She has a specially curled tentacle that she uses to roll the pollen into a neat ball after scraping it from the anther. She then tucks it under her tentacle and flies off to the next yucca flower.

The reason for this effort is that the female Yucca Moth is preparing to supply her young with food. The young Yucca Moth larvae eat developing yucca seeds. So the female Yucca Moth must ensure that seeds will develop from the flower where she deposits her eggs. She first packs the sticky pollen onto the stigma, the top of the carpel, which is a female flower part. This act ensures pollination and fertilization of the flower. Once this step is completed, the female moth moves to the base of the carpel, nearest to the part of the flower where the seeds will develop following fertilization, and lays her eggs. Her long ovipositor pierces the wall of the carpel such that the eggs are placed in the same cavity where the seeds will develop. Once the eggs are safely tucked inside the carpel base near the seeds, the female Yucca Moth heads off to collect another ball of pollen. She will continue to lay eggs in different flowers for as long as she has eggs, a period that may be as short as a few days or as long as a week. Inside the carpels, the seeds start to grow and so, too, do the Yucca Moth larvae. In time the fully developed larvae are ready to form pupae, but to do so they must leave the enlarged carpel base, now called a capsule, which they do by tunneling through the capsule wall and dropping to the ground. The larvae burrow into the soil and remain there, slowly developing into adult moths for a year or more, until the yucca plants are again in bloom.

The Yucca Moth cannot complete its reproductive cycle

without the chaparral yucca plant, and the chaparral yucca plant cannot reproduce without the Yucca Moth. This is a special example of what is called coevolution, where the biology of two species develops over time in such a way that each becomes adjusted to the other. The trick to this particular system is that not all the yucca seeds are eaten by the larvae because many more are produced than can be eaten. Pollination is made certain, and the plant pays the price of a few seeds to feed the moth larvae. Because of this special relationship, should anything happen to either species, the other would be doomed. Chaparral yuccas tend to flower simultaneously over large areas. That is why in certain years chaparral hillsides are thick with the conspicuous tall "candles" of yucca inflorescences. The mass flowering produces a favorable setting for the Yucca Moth reproduction and thus for the yucca plants too, but the mechanism for simultaneous flowering remains a mystery. Perhaps the moths themselves carry a signal to tell each generation of plants when to flower, or perhaps the plants produce a signal when the environment is favorable for moth reproduction.

After the yucca plant has flowered and died, it becomes host for a variety of insect species. The dried flowering stalk serves as the domicile for another insect, the California Carpenter Bee (*Xylocopa californica,* family Anthophoridae, order Hymenoptera). This huge, solitary, dark blue insect can be as much as an inch long. Despite the formidable appearance of this slow-flying bee, it is not aggressive. The female chews a tunnel into the stalk, where she lays about six eggs, each walled off in separate chambers made from sawdust chewings. The larvae feed on provisions of nectar and pollen brought to the nest by the female. The Black Yucca Weevil (*Scyphophorus yuccae,* family Curculionidae, order Coleoptera) bores as a larva into the basal stem of the dead yucca, and the adult feeds on stem sap. But most of the decomposition of dead yucca stalks is done by the Yucca Longhorn (*Tragidion armatum,* family Cerambycidae, order Coleoptera), a beautiful orange and black, one-and-one-half inch long beetle. The adult female feeds on the blossoms and stem sap of the flower stalk and then lays her eggs on the outside of the stem, so that the hatching larvae can bore into the drying fibers of the stalk. The larvae excavate long galleries through the pithy center of the stalks and may eat up to half of the lower stalk. Larvae pupate within these galleries and emerge as adults in late spring or summer in concert with the flowering of nearby yuccas.

Darkling Beetles (Order Coleoptera)

Darkling beetles are often seen late in the day ambling along paths in chaparral and oak woodlands. They belong to the genus *Eleodes* (family Tenebrionidae). There are approximately 100 species of darkling beetles in California, ranging in length from .4 to 1.5 inches, and it takes a specialist to tell one species from another. They have smooth, shiny black bodies with a pointed abdomen (pl. 70). These beetles are unable to fly, having traded flying wings for a fused upper body, thickened to protect it from the bites and pecks of predators, and well suited to retaining moisture in a dry environment. They are often noticed because of the peculiar habit of per-

Plate 70. A darkling beetle performing a threatening headstand.

forming a headstand when disturbed. This maneuver is meant to display the tip of the abdomen, where a noxious defensive fluid is emitted, to deter potential predators. This behavior explains the origin of two other common names for darkling beetles: stink bugs and circus beetles. This method of defense is apparently so effective that it is mimicked by other species of beetles that have a similar appearance but do not exude defensive chemicals.

Plate 71.
A tarantula hawk wasp sitting on flowers.

The Tarantula Hawk Wasp and the Tarantula Spider

Tarantula hawk wasps (*Pepsis* spp., family Pompilidae, order Hymenoptera) are good at coping with the scarcity of water and food in the chaparral. It is a very large wasp, a dark steely blue to black, and usually has contrasting orange red wings (pl. 71). It can be up to two inches long and is armed with a formidable stinger. The adult wasps actually eat only nectar; however, in order to provide for their young the adult females hunt the largest spider in the chaparral, tarantulas (*Aphonopelma* spp., family Theraphosidae, order Araneae). To the wasps the tarantulas represent a source of food and moisture for their young.

The life and death battle between tarantula hawk wasps and tarantulas is among the most dramatic struggles to be witnessed by a visitor to chaparral. A tarantula hawk wasp

Plate 72. A female California Tarantula *(Aphonopelma eutylenum)*. Despite its formidable appearance, this and other tarantulas of California are not dangerous to humans.

weighs less than .02 ounces, about the weight of a large paper-clip. Tarantula spiders, on the other hand, are hundreds of times larger than the wasps (pl. 72) and outweigh them many times over. The wasps can maneuver much better than the bulky tarantulas, and they generally win by flying around a spider and stinging it several times on the underside. However, the wasps do not always win. The spiders have strong jaws and a most formidable defensive posture. An unwary wasp can be quickly dismembered. It seems an unlikely match, but the victory is never secure and each battle has its own outcome.

In those cases where the wasp is the victor it claims its prize with great effort. The wasp has stung the tarantula only enough to immobilize it so that she can bring it to her nest, a hole dug in the ground. This can be a lengthy process since it may be several hundred yards to the wasp's nest. To move the huge tarantula the wasp beats its wings furiously while using its jaws to grip the spider's leg or antenna. It then drags the spider along in a series of jerking steps. Once the spider is stuffed into the nest the wasp may sting it again, but still does not kill it. The spider is in a comalike state where its basic metabolic processes continue at a low level. The wasp then lays her eggs on the tarantula, seals the burrow, and departs. The living spider is her solution to having fresh, juicy baby food for her larvae once the eggs hatch. The larvae feed on non-life-sustaining portions of the spider's body until they are near maturation, at which time they consume the remainder, killing the tarantula.

Trap Door Spiders, Ticks, and Scorpions (Class Arachnida)

Three types of arachnids—spiders, ticks, and scorpions— are often found in the chaparral. Arachnids are most easily distinguished from insects by counting their legs. Members of this class, including the tarantula previously described, have eight legs, while adult insects have six legs. The bodies of insects are divided into three parts: the head, the thorax from which the legs emerge, and the abdomen. A spider's body has two parts: the front section includes the head and legs, with the abdomen behind. The ecology of chaparral arachnids is poorly studied overall, but a few are well known. Trap door spiders are well adapted to life in the chaparral, including the inevitable fires. Ticks are also well adapted to chaparral, and to other plant communities in a wide variety of habitats across North America. They are among the likely invertebrates to be encountered by the chaparral adventurer, especially in spring, when a hike in the chaparral is most inviting. Scorpions are also surprisingly common but are seldom seen by people not looking for them because they are nocturnal.

Trap Door Spiders

Members of the order Araneae, trap door spiders (*Aliatypus* spp. and *Bothriocyrtum californicum,* family Ctenizidae) live in burrows in banks and hillsides in chaparral. They prefer sunny, south-facing slopes with grass or low herb cover, where they receive the sun's warmth early in the day. They are called trap door spiders because the door to their burrow is hinged on the uphill side and has beveled edges to fit the sides of the burrow opening tightly, much like a ship's hatch.

Hunting by trap door spiders is a sit and wait process, like that used by the Coast Horned Lizard. This is an energy saving process for the spiders, and one that keeps them from becoming food for someone else. Trap door spiders wait for the unwary beetle or cricket to wander past the burrow, pouncing on it and sinking in their fangs with a quick movement. They quickly drag the prey inside their burrow, where they suck the juices from its body. They usually keep one of their eight legs in the door of their burrow so they can disappear quickly if a spider-eating predator should come by looking for a quick meal. When they are not hunting, the spiders hold their well-camouflaged door shut with their fangs, because they too are a nice juicy package for a larger predator. This precaution is often insufficient protection, however. Skunks *(Mephitis mephitis* and *Spilogale putoris)*, in particular, are fond of trap door spiders, and they can quickly dig into the burrow to extract their prey.

Trap door spiders constantly remove dry soil from the burrow so that they can maintain high humidity and keep their body in contact with damp ground. The damp earth keeps temperatures down in warm weather and allows these rather large spiders to breathe without drying out. They can lose water rapidly if forced to be active when it is too hot and dry.

Trap door spiders are adapted to the periodic fires of the chaparral. During a fire their carefully constructed door burns off, exposing the opening to the burrow. However, the spiders survive because the vertically dug burrows are six to 11 inches deep, providing an environment that may remain cool enough for them to survive. Another advantage of the burrow is that smoke from the fire does not tend to be pulled down into the narrow, vertical shaft. A new door is fashioned by the spider within 24 hours of a fire. On the other hand trap door spiders may move to other locations after a fire. It is common to see the exposed upper portions of trap door spider burrows, without doors, sticking up like little chimneys on slopes during the months following fire. The upper layer of soil often erodes from these bare slopes, leaving the exposed necks of empty, silk-lined spider burrows standing well above the surface of the ground.

Ticks (Order Acari)

A person hiking through the dense vegetation of chaparral often picks up an unwanted companion, a tick. This is particularly true in winter and spring. These little arachnids sit on the edges and tips of vegetation, from which they drop onto passing animals when the plant is disturbed. Ticks are external parasites that extract blood by temporarily but firmly attaching themselves to the skin of the host animal, where they extract a blood meal. Fortunately, ticks usually take several hours to firmly attach, so you have time to discover and remove them before any harm is done. They are known to transmit a number of serious diseases such as tularemia (hunter's disease) and Lyme disease. An unfed tick is less than one-tenth inch in diameter, with eight grasping legs and often a surprisingly hard body surface (exoskeleton), but it can quadruple in size after a good meal, coming to resemble a small blimp with legs. Four species of ticks will bite humans and are commonly found in chaparral. Experienced hikers know that it is wise to inspect the entire surface of the body at the end of a day spent in dense vegetation in order to find and remove ticks. These little parasites prefer parts of the body that are snugly protected from exposure, such as the hairline at the nape of the neck, and places where clothing is in close contact with the skin—socks, underwear, and the like.

Scorpions (Order Scorpionida)

Chaparral scorpions vary in size from less than an inch to 2.5 inches from head to extended tail. They emerge from their shelters only at night. They live under rocks, in rotting wood, and in burrows they excavate. None of the species found in chaparral areas are life threatening, but they can deliver a bee-sting level of pain. The large and common Burrowing Scorpion (*Anuroctonus phaiodactylus,* family Vejovidae) is found throughout chaparral, and its burrows often line the sides of trails cut into slopes. It can be recognized by its stout body up to 2.5 inches long, and bulky, reddish brown claws. Unlike the burrows of spiders and insects, which usually have round openings, scorpion burrows are almond shaped, reflecting the flattened body shape of their owners. Because scorpions keep their burrows cleared of soil, they can easily be discovered from the small pile of fine sand grains below the lip of the opening. At night scorpions lie in wait at the burrow entrance, where they use their powerful claws to grasp and hold insects and other small invertebrates. One of the peculiarities of scorpions is that their bodies fluoresce when exposed to ultraviolet light. Scorpion collectors often search for them at night with portable black lights, under which they glow various shades of blue, purple, or green.

Other Chaparral Insects

Of course, many other insects occur in chaparral. One group that attracts some negative attention is the canyon flies (*Fannia* spp., family Muscidae, order Diptera). These small, gray flies are only about an eighth of an inch long but have the aggravating habit of landing on exposed skin and crawling into the eyes, nose, ears, or mouth in search of moisture. Although they do not bite, their persistence and sheer numbers can make them distracting and annoying.

A number of moths also live in the chaparral. Although they are generally nocturnal they may occasionally be seen at dawn and dusk. One notable species, the Ceanothus Silk Moth (*Hyalophora euryalus,* family Saturniidae, order Lepidoptera) is a beautiful chestnut brown color with a wingspan up to five inches wide (pl. 73). The sausage-shaped green lar-

Plate 73.
A Ceanothus
Silk Moth.

vae can be up to four inches long, and the body is studded with yellow tubercules (bumps).

Along with moths, a number of butterflies, bees, and beetles are also found in chaparral, many of which are important for pollinating the flowers of the shrubs. Relatively little is known about these relationships, but it is likely that some are every bit as complex as that known for Yucca Moths. One group, that of solitary bees, is particularly numerous in chaparral, and as a rule each species feeds on only one plant. A survey of chaparral at Pinnacles National Monument in the Central Coast Ranges found 410 species of bees, a higher number than known from any other ecological community in North America, and one-tenth of all the bee species known in the United States.

It would be reasonable to suppose that when chaparral is swept clean by an intense wildfire, all insects would be temporarily exterminated. This is not the case. Many species of chaparral insects can survive fires by living underground, inside burls or trunks and stems of larger plants, or under rocks and other protective elements. These species may be stimulated to activity by the heat of the fire, flying to fire or freshly burned plants to reproduce, as is the case with the fire beetles (*Melanophila* spp.) described in chapter 3. A study of insects in the months following an intense wildfire in chamise-ceanothus chaparral in the San Gabriel Mountains showed a peak of insect species and abundance that declined over the following three years. These insects may have survived the November fire in recesses beneath the ground, or they may have quickly crossed the distance of a mile or so between the burned study area and the nearest unburned chaparral. The large number of plant species that quickly appeared and flourished after the fire may have contributed to the peak in insect diversity. In general, most species of insects are closely tied to the plants upon which they live. The flush of insects after fire may explain the abundance of insect-eating animals in chaparral at this time. For example, the total number of bird species observed in recently burned chaparral is about the same as in mature chaparral, even though the particular species composition is somewhat different before and after a fire.

THE BEAUTY AND UTILITY of chaparral is tempered by inherent problems and dangers. The gentle plain of Santa Barbara to the Sierra foothills and the bucolic canyons of the Santa Monica Mountains all bear testimony to the power of chaparral wildfires to invade our cities and suburbs (pl. 74). In the aftermath of a chaparral wildfire on steep terrain, the bare steep slopes become unstable. Heavy winter rains can quickly create a massive slug of water, mud, and debris peppered with boulders that surges uncontrollably down slopes and canyons into settled areas. These viscous flash floods can materialize suddenly and with no obvious warning, so that people in their path have almost no time to prepare or to flee.

Chaparral wildfires that destroy property, and sometimes take human lives, are not acts of God that abruptly and capriciously materialize, striking down the unsuspecting. They are quite comprehensible and predictable, in the same sense that floods are predictable for people living on the flood plains of rivers in other parts of the United States. Dwelling on the 20-year flood plain does not mean that a house will experience a flood at 20-year intervals, or ever, within the tenure of a particular occupant, but rather that the chances of a flood in a given year are one in 20. Similarly, living in chaparral where fire has come three times in the past century does not mean the next fire will come precisely 33 years after the last. After the shrubs reach the point where they contain enough fuel to readily burn in dry weather, each passing year without fire slightly increases the potential flammability and probability of fire the following year. So the risk of flood or fire is more like that found in the actuarial tables of life insurance companies, where the older one becomes the greater becomes the probability of death. People sometimes conclude that a particular location in chaparral is safe from fire or flood because it has been many years, perhaps longer than a human lifespan, since disaster has struck that particular place. The same wishful thinking permits some people in an area with high seismic risk to live or work in a building that has stood for many years

Plate 74. A house located at the top of a canyon burns during the Malibu fire of 1985.

even though it would be quite dangerous in a strong earthquake. Upon a landscape shaped by episodic fire or flooding it is wishful thinking to assume that these recurring events will cease. It is just a matter of time until disaster strikes.

Even though chaparral wildfires are inevitable, and under some circumstances may even be desirable, the first concern of residents and public officials alike is protecting lives and property. For about 100 years this concern has been addressed with policies of fire prevention and of extinguishing fires as quickly as possible when they do occur. Necessary and expedient as these policies may be, they are not a complete and satisfactory way to deal with wildfires in chaparral. Moreover, other management goals, such as fuel reduction, watershed improvement, and wildlife habitat enhancement, require different responses. Fire is inevitable wherever chaparral occurs. Any policy that fails to recognize this fact is fundamentally flawed.

Prescribed Fire

One tactic for reducing the danger of living near chaparral is to use fire to reduce the volume of fuel in critical areas. This is sometimes done by deliberately burning patches of chaparral in a carefully controlled fashion. The idea of intentional burning—prescribed fire—is to buffer urban areas adjacent to cities with chaparral that is young enough to retard or stop a wildfire approaching from distant, older, and more flammable chaparral. Prescribed fire has gradually assumed an important role in the control of chaparral wildfires as managers have realized how useful it can be in the long term, resulting in fewer large wildfires, less loss of life and property, and lower expense. Prescribed fire may be used to burn patches of hundreds to thousands of acres, or it may be employed to create smaller natural clearings in the potential paths of destructive wildfires. The photograph at the beginning of chapter 3 shows a large prescribed fire done for research purposes during summer burning conditions. Since fire after fire often follow a similar path through mountain passes and along hillsides, it is possible to predict the most likely places for large fires to occur. For example, patterns of Santa Ana winds and urbanization mingled with large and continuous stands of chaparral have pushed destructive wildfires along the same avenues of chaparral time after time. One way to break these patterns is by removing wide swaths of mature chaparral right across the potential fire paths. Prescribed fires can sometimes provide this kind of clearing so that fires can be more easily and safely blocked before they move into areas where lives and property are at stake.

The shift from the failed policy of suppressing all chaparral wildfires to using prescribed fire as a management tool has not been simple or easy, and it still is not. A number of difficulties and limitations will always make the widespread use of prescribed fire in chaparral and elsewhere problematic. The possibility of an escaped prescribed fire will always be something that causes managers to think long and hard before authorizing ignition. Fire managers know how to predict the general behavior of burning chaparral by using careful measurements of weather conditions, characteristics of living and dead fuel in the chaparral itself, and topography. All of these variables can be entered into computer models that help to predict how fire ignited at a particular time and place will burn and spread. The accuracy of these fire predictions is only as good as the weather forecasts upon which they are based, especially the information on wind and relative humidity. The areas to be burned are carefully prepared in advance, so that various kinds of natural and artificial barriers will prevent a prescribed fire from spreading outside of the intended perimeter. The prescription for a particular fire specifies maximum and minimum acceptable levels of temperature, relative humidity, wind direction and speed, and fuel moisture. If there are any deviations from these prescribed conditions, the fire will not be attempted. The goal is to ignite chaparral when it is dry enough to burn, and for flames to gently move through the intended fuel. It must be possible to contain the fire under these conditions. I (R.Q.) once observed the ignition of a management fire when prescribed weather conditions were met, but shortly after ignition the wind direction shifted, which doubled the level of relative humidity in a matter of minutes. The chaparral quickly became too damp and the fire went out without spreading at all.

The greatest risk in prescribed burning is a sudden and unexpected change in the weather after the fire has been ignited, such that it causes the fire to become unmanageable. Under the right conditions, a prescribed fire may escape despite the most careful plans. On one unlucky day in spring,

Santa Ana winds that had not been predicted suddenly developed in southern California, which caused several prescribed chaparral fires that were burning in various places at the time to get away. The risk of escape rightly causes decision makers to err on the side of caution. An escaped prescribed fire will be remembered far longer than the many successful and safe burns, particularly if it causes damage. Months and even years of careful planning can come to naught in the face of a dry spring or persistent winds. Often there is nothing to do but wait another year and hope for more favorable weather. These difficulties notwithstanding, prescribed fire holds out a useful and relatively natural method for reducing the chaparral fire hazard in selected areas. It can be a useful tool for small stands of chaparral near urban areas, but it will never be practical in most chaparral because prescribed fires are expensive to plan and execute, particularly in rugged areas with limited access.

Public perceptions and opinions about prescribed fire are another important ingredient in its usefulness. Most people are not accustomed to seeing chaparral or any natural vegetation deliberately burned. Without some explanation they may be disapproving or even alarmed at the prospect. It is especially important that nearby residents, who may treasure the scenic value of mature chaparral, understand the purpose and necessity of burning parts of it in the interest of public safety. There have been instances where federal, state, and local agencies have found their efforts to reduce chaparral fuels with prescribed fire thwarted by people who did not want vegetation deliberately burned on or near their properties and communities. Smoke is a related community and regulatory concern. Much of California's chaparral lies on hillsides adjacent to valleys with polluted air, so that fire managers are permitted to burn only on days when smoke from the fire will not add to local smog or cause discomfort to people living nearby. The requirements of laws protecting endangered species of plants and animals can further complicate planning of prescribed fires, because no action can be undertaken that might harm such species.

Prescribed fires are generally ignited in late spring, after fuel moisture has dropped sufficiently for chaparral shrubs to burn but before the vegetation dries out sufficiently to produce such an intense fire that might be difficult to control. Fire at this time may cause more damage to soil, seeds, and reproducing animals than fires in other seasons. Studies have shown that spring fires cause different and more severe changes in many biological processes in the soil, especially if soil moisture is present, than do fires over dry summer soils. Spring burning can disrupt or terminate reproduction of birds at the peak of the breeding season. Populations of chaparral plants can lose an entire year's reproductive effort if flowers or fruits are present on the plant when it is burned, as often is the case in spring. These multiple impacts on ecological processes caused by spring burning can retard the recovery of chaparral vegetation and may ultimately shift the species composition. Prescribed burning of chaparral can be done in winter, the other season when wildfire is unlikely. However, low temperatures and high fuel moisture often make it difficult to impossible to burn stands of living chaparral shrubs at this time. Winter burning can be used to safely remove piles of dead vegetation during weather when surrounding shrubs are virtually incombustible.

Sometimes firefighters will deliberately set a prescribed fire, called a backfire, in the path of a spreading chaparral wildfire. Their goal is to deprive the uncontrolled fire of fuel when it reaches the place where the backfire has already consumed the vegetation. This is done in places where a barrier, such as a road or clearing, can prevent the backfire from spreading in the same direction as the wildfire, which is usually the way the wind is blowing. This is done only when few other options exist for stopping a wildfire, because there is always the danger that the backfire will spread and become part of the fire it is intended to block.

Fuel Reduction and Fuel Breaks

The fuel contained in chaparral shrubs can be reduced on a local scale by a number of methods. Bulldozers, often pushing or pulling large implements, can roll, chop, crush, plow, or otherwise reduce shrubs to small pieces that lie upon or within the soil. In this state they are incapable of carrying an intense fire. Most of these mechanical methods cause severe soil disturbance and may kill the root systems of the shrubs and other perennial vegetation. In very small areas, hand labor similar to that used to build fire lines can be employed to thin or cut shrubs. This method is expensive because it is labor intensive. Shrubs or their stumps may be killed through treatment with herbicides. This method must be used with great care because of the potential damage from chemicals to the environment and other organisms. If intact dead shrubs are left by any of these methods, they must be removed by bonfires or in some other way before the next fire season, because they are far more flammable dead than alive.

Goats are an increasingly popular tool for reducing the volume of chaparral shrubs in small areas. These animals will eat almost any kind of chaparral vegetation, and the amount and type of plant material they remove can be governed by herding. Goats can be cost-effective because as they reduce fuels they are being fed free of charge, all the while producing valuable products such as Angora wool. Goats, unlike cattle, must be closely tended because they are sensitive to cold or wet weather and are subject to predation by Coyotes *(Canis latrans)* and Mountain Lions *(Puma concolor)*.

Under extreme conditions chaparral wildfires can spread furiously in any direction the wind blows, fed by voracious consumption of the shrubs. One way to thwart the spread of a wildfire is to deprive it of additional fuel. In some places this is done by creating long roadlike gaps in otherwise continuous chaparral by removing most or all of the shrubs. These barriers to fire, called fuel breaks, are often constructed in broad swaths between urban areas and expanses of chaparral in adjacent wildlands (pl. 75) using a selection of the methods described above. A progressing wildfire can sometimes, but not always, be safely stopped at a fuel break. As described later in this chapter, under extremely dry and windy conditions chaparral wildfires can spread furiously, and flying embers and firebrands can easily carry fires across fuel breaks and even much wider barriers. Fuel breaks 200 to 300 feet wide are best employed as part of an integrated fire management plan that includes other means of fuel reduction as well. As shown in pl. 75, fuel breaks are often placed along ridge tops because these are places where fires that have raced up slopes under the propulsive force of convection tend to naturally slow down. Ridge tops are also places in rough terrain where firefighting roads can be most easily constructed.

The edges of fuel breaks are often deliberately made irregular, because this produces a more pleasing and natural appearance. The look of a fuel break can be further softened,

Plate 75. Swaths of cleared land adjacent to an urban area form fuel breaks in chaparral, in the San Rafael Hills, Glendale, California.

without losing its effectiveness, by leaving small patches of chaparral in especially wide places. Even trees can safely remain in some places as long as all lower and dead limbs are removed so that a ground fire cannot climb into the tree canopy. After shrubs have been removed, fuel breaks are usually planted with another type of vegetation that does not pose the same degree of fire hazard as chaparral, such as perennial grass or other plants that grow close to the ground. This cover vegetation reduces the danger of erosion and also improves the appearance and wildlife value of fuel breaks. The clearings of fuel breaks can become inadvertent avenues for invasion by plants that grow well in places where the soil and natural vegetation have been disturbed and reduced. These plants are usually nonnative species, and they can spread into uncleared chaparral areas, especially after fire.

Fuel breaks are expensive to construct and require regular maintenance in order to retain their effectiveness. Their cost is such that they are unlikely to be placed throughout chaparral or around every settlement or dwelling near chaparral, but their strategic use can be of enormous benefit to firefighters. Chaparral wildfires have been slowed and stopped by fuel breaks, and attendant damage greatly reduced, in all parts of California. In some cases this has broken a historical pattern of wildfire spread, reducing potentially large fires to much smaller dimensions. In principle, the money fuel breaks save by reducing the cost and damage of a threatening wildfire can be greater than the cost of the fuel break itself, in addition to the benefits of reduced danger and human suffering.

Artificial Seeding of Burns

Seed of annual grasses, especially Italian ryegrass (*Lolium multiflorum*), has often been scattered from aircraft on burned chaparral slopes. This practice is an attempt to establish plant cover quickly on a denuded landscape, with the expectation that there will be less soil erosion from land covered with plants than from bare soil. While ryegrass seeding on chaparral burns has gone on since the 1940s it is of questionable value. Most chaparral ecologists and environmentalists oppose the practice, having concluded from research that over the long run the practice may actually make erosion worse. In addition, it may disrupt the native plant community.

It might seem obvious that the faster a burned slope is covered with well-rooted plants, the less erosion when rain falls. However, the vigorous growth of ryegrass requires gentle and evenly spaced rains throughout winter—conditions seldom met, particularly in southern California and the Sierra Nevada. It is more likely that the majority of rain will fall, with attendant erosion, before ryegrass seedlings become well established. If rainfall is spread out through winter so that ryegrass grows well, then the spreading grass can compete with the growth of seedlings of chaparral shrubs. Moreover, when the annual ryegrass dies the following summer it can provide a fine, continuous fuel that could easily carry a second fire across the same landscape. A reburn at this time would kill all chaparral shrub seedlings and injure or kill resprouting shrubs as well. An additional problem with ryegrass planting is that the vigorously growing grass plants may out-compete the native fire annuals and other herbaceous plants. This produces a cascading negative effect, because if the native plants do not survive to flower and set seed, then there will be fewer seeds in the soil to repopulate the area with natural chaparral plants after the next fire. Once set in motion, this process, reinforced by subsequent fires, could eventually lead to the complete elimination of native plant species including the ground-holding shrubs that control erosion naturally.

Seeding with nonnative plants after chaparral wildfires continues to be practiced for two reasons, despite all the evidence that it is an ineffective practice for both erosion control and for conservation of the native flora. First, it is something managers know how to do and traditionally it has had public support. It is "something" that can be done. Second, even with all its potential problems, it is not possible to predict exactly what will happen in any given instance after ryegrass is sprinkled from the sky onto a blackened chaparral hillside. Seeding aside, it is not possible to predict exactly what species of plants will come up in a given location, or when or in what numbers, and harder still to conduct any practical research that would provide a universal set of management recommendations. Many ecologists and environmentalists have suggested that no further artificial seeding of any kind be done in burned chaparral. A compromise would be to use the seeds of native chaparral plants in place of the exotic grasses. In recent years some public management agencies have been directed to do so when practical. This is occasionally done, but not often. Very large quantities of seeds from native annuals are difficult to obtain on short notice, and they are expensive. Furthermore we have no information on how well native plants might germinate and grow when introduced in this way, much less whether they might retard erosion any more effectively than naturally occurring seedlings. Equally important, there is no assurance that large numbers of introduced seedlings of native plants would not destabilize populations of seedlings of other species of plants natural to the site. It is clear that artificial seeding of any kind has the potential to produce results that are different from what occurs in naturally recovering chaparral. The long-term consequences of these changes are unknown and unpredictable.

Fire Causes Its Own Weather

The peak fire season in California chaparral is correlated with the end of the rainless summer, which can bring hot, dry weather and erratic winds. However, these are not the only weather conditions that cause the spread of fire. Once a wildfire builds to sufficient intensity and size, it can create its own weather, a firestorm, causing it to build upon itself and gather the potential to start new fires. This occurs because air heated rapidly at ground level by the fire creates strong updrafts as it rises. The internal turbulence from the smoky, hot air traveling upward in enormous columns can rapidly draw in more air along the ground, and this wind erratically pushes the fire in new directions. In addition, this upward movement of hot air pulls along embers, firebrands, ash, and smoke as it rises. Once a hillside is on fire, the masses of surging, fiery debris and smoke can be propelled upward to as high as 20,000 feet! There the clouds begin to cool. Meanwhile these churning masses of hot air and debris are moving in the air currents hundreds to thousands of feet above the fire. These drifting fire clouds sometimes rain burning pieces of wood, ashes, and cinders as far as several miles from the original source. The fire's own internal weather system also creates gusty windstorms over a wide area. Consequently, the original fire is driven and spread erratically in the face of the unpredictable forces of the firestorm. It is almost impossible to contain a wildfire in chaparral when it is driven by strong winds. The winds spread the fire clouds and fan the smoldering bits inside them so that once dropped to the ground, they are likely to start satellite fires. As these secondary fires build up, the entire process can be repeated. When it is very windy, as is the case with Santa Anas and other such winds, little can be done to prevent fires driven by the wind from moving unimpeded through chaparral. Aircraft, an important tool in fighting wildfires, may be restricted from flying close to the fire because of high wind velocities or dense smoke that obscures vision.

Houses built on steep slopes covered with chaparral (pl. 76) can be particularly vulnerable in wildfires, because the updrafts of fire bring the scorching air and burning debris straight up the hill to the house. Structures in canyons may be in equal peril during severe fires, because the canyons can act like wind tunnels, funneling the wind-driven flames at hurri-

Plate 76. These houses, surrounded by chaparral, are built on steep slopes with only one access road.

cane speed. Under severe fire conditions almost nothing can be done to save structures in vulnerable locations. These fearsome blazes cannot be suppressed or even safely approached by firefighters and equipment, either on the ground or in the air.

Geographic Risk

There is a degree of geographic predictability about chaparral wildfires and subsequent floods. Certain areas, because of the age of the shrubs, topography, wind patterns, and other factors, are far more likely to experience wildfire than are other areas with lesser or fewer risk factors. Similarly, since water and floods follow the path of least resistance downhill, it is not difficult for an experienced hydrologist to predict how much and where water and debris will be transported when heavy rains fall on recently burned chaparral slopes.

In the 1970s I (R.Q.) used to visit an ecologist friend who lived in the chaparral. His house was tucked into a canyon in the Santa Monica Mountains, not far from the urban bustle of Sunset Boulevard. He lived with his family in a rented house within a California state park. The house was constructed mostly of wood, including a rustic wood shake roof, and it was so intimately involved with chaparral that he ran sensor wires out of the living room and into the adjacent shrubs where he was monitoring the behavior of wood rats. The chaparral was quite old and very thick. The house was accessible only by a single-lane, winding road that made its way up the canyon. We would sit in his living room and discuss what he and his family would do when—not if, but when—the chaparral wildfire came. There was no question that the house would burn if the flames arrived during extreme fire conditions. The house was extremely flammable, and there was no way to safely defend it. Close to the house there was a sort of concrete bunker set in the bank above the stream. It had been constructed for another purpose long ago, but for him it was to be a fire shelter for the inevitable disaster. Three years later that house did burn down, as did about 150 others in one of the periodic wildfires that sweep through the chaparral-clad canyons of the Santa Monicas.

To the unschooled eye, one chaparral area may appear no more or less dangerous than another, but that is not always true. This is why the California Department of Forestry and Fire Protection (2002) has prepared a *Fire Hazard Zoning Field Guide* that categorizes specific areas of the state according to the wildland fire danger they face. According to this guide, in the year 2000 more than 100 cities and counties contained "very high fire hazard severity zones." About two-thirds of these were in southern California, and most had chaparral as the underlying cause of hazard. This zoning system allows communities to identify risks to public safety and to take steps to protect themselves, and many towns are doing so.

The most destructive wildfire of its time spread from natural vegetation through an urban area of the Berkeley Hills in 1923. Fueled by a mixture of naturalized eucalyptus trees, grasses, and shrubs, it destroyed 584 structures. A half-century later, fire managers predicted that a very similar catastrophe would happen again because the conditions of wildland fuels and patterns of adjacent urbanization had not improved since 1923. On the contrary, they had become worse. The prediction came true in 1991 when the Oakland Hills fire destroyed nearly 3,000 homes and cost 25 lives, making it the most destructive wildfire to ever occur in California and ranking it as one of the worst natural disasters to ever befall the state.

Floods

In those numerous parts of California where chaparral is located on steep hillsides above urban and agricultural areas, the inevitable fires sometimes bring the subsequent risk of flooding. These are not floods in the usual sense of the word but are more like sudden assaults by muddy slurries that can wreak havoc in a matter of minutes. The relationship between fire and flood is direct but not obvious, and it is often misunderstood or overlooked until it happens. The following sections explain the connection between the two.

Mature Chaparral Acts as a Sponge

The continuous cover of mature chaparral shrubs acts like a vast sponge. Roots hold the soil in place, and the shrubs collect rainfall in such a way that water is gradually and continuously released from the mountains. The tangle of interlocking branches and leaves breaks up and deflects the impact of falling raindrops on the ground, so that the water is scattered and slowed, and droplets are eased into the soil. The layer of leaves, twigs, and other debris dropped from the shrubs onto the soil sets up another series of barriers to the downhill flow of water. The matrix of litter creates tiny dams and obstacles to trickling water, dividing and diverting it into innumerable and circuitous pathways. Robbed of most of its downward velocity and power, the weakened flow seeps into the soil rather than moving down across the surface. While surface water from chaparral-covered hillsides may slowly flow across valleys and out into the ocean, most of the flow is below the surface, collecting in underground basins called aquifers, where it can be tapped long after the rainy season has ended. Above or below the surface, the water can be retrieved, stored, and diverted for future use by humans. Aquifers and local surface water were the sole sources of water throughout California before it was possible to import water from distant sources. Local water originating from chaparral watersheds continues to provide an economical source of much of the water used by people across the arid valleys and coastal plains of central and southern California.

Soils and Heavy Rain

Trouble begins when heavy rain falls upon steep chaparral slopes that have been recently laid bare by fire. Denuded of cover, these once benign and efficient watersheds lose most of their capacity to absorb rainfall. Keen observers long ago noticed that when people walked across the soil of recently burned chaparral vegetation after moderate rainfall, that their boots would turn up dusty footprints. Unlike ordinary soil, these burned chaparral soils were wetted by rain only very near the surface, leaving everything below powder dry. In effect, the fire had somehow waterproofed the soil so that water that could not be accommodated by the top inch or two of soil was forced to run off across the surface. Scientists discovered that the mature chaparral shrubs produce organic compounds with waxlike properties that are deposited on the surface of the soil from litter that falls from the plants. These durable compounds are gradually incorporated into the uppermost layer of soil as the litter decays. When a fire occurs, the waxy substances are heated into gases and driven down beneath the surface of the soil to a depth where cooler temperatures allow them to recondense. There they are concentrated around particles of minerals, creating a barrier to the passage of water as effective as a plastic sheet (fig. 56). Soil scientists call this a hard-to-wet soil.

The extreme heat of many chaparral wildfires, measured in excess of 1,000 degrees F at the surface, incinerates all the organic matter in the upper portion of the soil. This organic component, called humus, binds the mineral particles together like mortar between stones in a wall, and at the same time greatly increases the capacity of the soil to absorb and retain water. Deprived of humus, intensely burned soils lose their internal cohesiveness, becoming as loose and powdery as fine sand. So the burned chaparral hillsides are covered

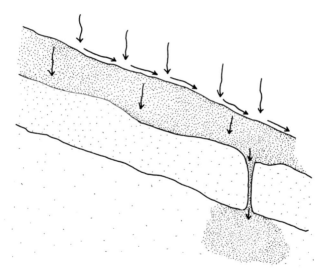

Figure 56. Hard-to-wet soil on a chaparral slope. A layer of water-repellent soil develops below the surface after a chaparral fire. While precipitation wets the surface layer of soil, it can penetrate more deeply only through occasional gaps (lower right) in this water-repellant layer. Paths of water are indicated by arrows, and wetted soil is darkly shaded.

with loose, unstable material with a limited capacity to absorb water, resting upon a layer that actually repels water. Compounding this precarious condition of the soil is the lack of plant cover to mitigate the effect of exposure to the full force of the elements.

Moderate or heavy rain soon saturates the uppermost layer of loose material, which begins to be carried downhill with the water. The heavy, soggy material on the surface slips easily down the surface of the hard-to-wet layer, like wet leaves on a steep metal roof. The water collects into small muddy rivulets, roughly parallel to one another, and cuts a series of rills where the soil has been carried away. The displaced material collects at the bottom of slopes and canyons.

Floodplains

The geology and climate of California have conspired to place many of the most desirable and useful parts of the state in areas subject to periodic flooding. As long as there have been humans in California, they have been attracted to the relatively level places around the aprons of hills and mountains, and the valleys between mountain ranges. These places generally have productive soils and a gentle climate, and when water is nearby, they are eminently suitable for agriculture and settlement. From the foothills of the Great Central Valley to the fertile Salinas Valley, south to the productive Oxnard Plain and teeming valleys of the Los Angeles megalopolis, people are living and working on floodplains, places where over millennia periodic outflows of water and debris from nearby mountains have shaped the landscape.

If all the material carried off slopes by gravity or water were to accumulate and lie at the bottom of hills and canyons forever, neither it, nor the condition of the chaparral above it, would matter. It takes a large flush of water, delivered suddenly, to move the debris out across the valleys. The floods from burned chaparral watersheds are not like those of the Midwest or those that used to inundate parts of the Sacramento Valley. In both of these cases, water slowly and inexorably spreads across gently sloping floodplains as a result of events far upstream. There is time to move people and animals to high ground and prepare for the flood's onslaught. Severe chaparral floods are quite sudden and unpredictable, not gradual. They are akin to the flash floods of Southwestern deserts, which turn previously bone dry washes and river beds into a sea of raging torrents that sweeps away all in its path. The imminent arrival of these muddy floods in the chaparral is sometimes announced by a roaring sound, with the volume and quality of an approaching fast freight train. And there is no more time to escape than to move off the railroad tracks.

A burst of rain near the end of a large storm is the least predictable antecedent of catastrophic flooding from burned chaparral watersheds. This sudden event is caused by a churning at the top of the storm's body, which quickly cools air that is already saturated with water. This turbulent force aloft quickly drives much of the moisture out of the air over one spot, as if wringing out a huge overhead sponge. The physics of this phenomenon is similar to that which generates violent thunderstorms in other climates. Both are relatively brief, localized, quite intense, and unpredictable. They are sufficiently rare as to be unknown to most Californians living in coastal valleys. The downpours can occur upon watersheds covered by chaparral at the middle elevations of mountain slopes that face the Pacific. A burst of rainfall that caused the La Crescenta flood, described below, lasted about 15 minutes, but it came on the heels of a 12 inch drenching. That particular downpour delivered more rainfall in minutes than a typical large Pacific storm produces in 24 to 48 hours. As described in chapter 2, rainfall events like these are unusual, but they are not unprecedented, freakish occurrences. As the pouring rainfall collects at the surface, it gathers a torrent of propulsive force as it is suddenly forced through steep and narrow canyons. The action of this water can grind a house to splinters and tumble crushed vehicles onto roofs in less time than it takes to read this paragraph. Canyons and dry water-courses can be quickly filled with moving slurries of water, mud, boulders, and other debris. These debris flows can scour out loose material that has accumulated in canyon bottoms for decades and then rush at the valleys below with astounding force, velocity, and volume—brown flash floods complete with rolling battering rams in the form of boulders.

The occasional very wet winters that bring floods to California have been a regular feature of the mediterranean climate for thousands of years. We have direct records of these rainy winters since the arrival of the first Europeans who began recording storm events. We have indirect evidence as well for millennia deeper into the past from patterns of flood deposits. Since rainfall records have been kept, we see a number of extreme years. For example the rains of 1862, which continued unbroken for 30 consecutive days, created a lake in the Sacramento Valley that was said to be the size, but fortunately not the depth, of Lake Michigan, leaving the new state capital under water. In another instance, in 1938, a very large flood struck southern California. The flow of the Los Angeles River at its mouth during this flood was greater than the average flow of the Mississippi River at St. Louis. That regional flood killed 119 people, and metropolitan Los Angeles had so much flood damage and standing water that road access to the outside world was cut off for several days. Few years are "average," but fortunately very few years are so extreme (fig. 4).

It is ironic that mediterranean California, famous for its dry climate, very occasionally experiences rainfall of an intensity seldom experienced in much wetter climates. Even with the steep mountains, and urbanization thoughtlessly placed on the floodplains and in the natural drainages below, it is difficult for people to understand and prepare for an event so infrequent that it is outside the experience of the average resident. Intense rainfall does not occur every winter, or even in every unusually wet winter. Concentration of the rain over a short period of a few weeks, typically caused by a cavalcade of

Pacific storms in close succession, produces floods. This pattern is often associated with El Niño conditions that periodically develop in the southern Pacific Ocean.

The Human Reaction to the Variable Climate of Chaparral: The La Crescenta Floods

On New Year's Eve 1933, it had been raining hard for 14 hours in southern California. Late evening parties were interrupted in the La Crescenta Valley north of Los Angeles as the power failed. A sudden burst of rainfall was so intense that water trickled inside buildings through tightly shut windows. Residents inside darkened homes heard an ominous rumble to the north, where the San Gabriel Mountains were disgorging a torrent of debris-laden water from several steep canyons. As the wall of debris cascaded down poorly defined watercourses and spread out across an area of three square miles on the valley floor, the noise of clattering boulders became so loud that conversations inside tightly closed houses became impossible. In 13 minutes the flood killed 49 people and destroyed or damaged 598 homes. Many of the homes were buried to the rooftops, and bodies of some flood victims were never found. They may have ended up in the Pacific Ocean, 30 miles away.

The chain of events leading to this tragedy began with a 5,000 acre wildfire that cleared chaparral from the steep mountain slopes above the valley the month before the storm. Rainfall on New Year's Eve was extraordinarily intense, depositing 12 inches on the bare mountain slopes. Just before midnight the trailing edge of the storm, accompanied by an upper air disturbance, passed across the valley foothills. The disturbance set off a violent downburst of rainfall that lasted about a quarter of an hour. This brief downpour, on saturated soil, set a sheet of water and debris in motion. The burned hillsides were quite steep, and the equally steep canyons held three decades of debris accumulated since the previous floods in 1914. As the rush of water and mud concentrated in these canyons, its velocity and force was amplified. Boulders as large as automobiles were sent tumbling and crashing against one another, driven by the torrent. One of these car-sized boulders was deposited in the middle of Foothill Boulevard more than two miles from the base of the mountains. It was later necessary to blast it into smaller pieces that could be lifted and hauled away.

This storm demonstrated that the La Crescenta Valley could be a hazardous place to live, but this had not been apparent until the unusual combination of a large chaparral wildfire, closely followed by prolonged and occasionally intense rainfall, brought the mountainsides crashing down. To insure that such a disaster would not occur again, natural watercourses were replaced with a system of linear concrete channels to divert water from canyons and slopes so that it would run quickly to the ocean. Debris basins were built in many places at the base of the mountains to confine the boulders, mud, and other materials near the point of origin, so that these materials would not cause damage and obstructions farther downstream. Debris basins look like dams, but they are designed with a different purpose. They stop floodwater long enough for the suspended materials to settle out, and then allow the cleared water to flow out into a flood control channel (pl. 77). When the impounded material fills the debris basin it must be removed by heavy equipment before the next storm deposits more. The La Crescenta system was specifically designed to protect the valley areas from future chaparral fire-flood disasters. Restraint of nature under the engineer's hand was, and is, widely and successfully applied for flood control in the valleys and canyons of California, but this solution is by no means foolproof.

Floods and debris flows still emanate from chaparral, especially after fire, but not on the catastrophic scale that would occur without elaborate flood control works. In 1975 the very same stretch of chaparral above La Crescenta that had burned in 1933 burned again, racing across an area 10 miles wide and 18 miles long in three days, while destroying eight valley

Plate 77. Aerial view of the debris basin at the mouth of San Antonio Canyon, San Gabriel Mountains. The diagonal gray band across the upper right is a berm that prevents water and debris from flowing out of the mouth of the canyon. The mountains at the top left and below are entirely denuded of chaparral by a recent wildfire. The green at the center is unburned agricultural land, and the light area at the center is a mining operation that processes rocks, sand, and gravel carried to the basin by periodic flooding. The floodplain below the basin is heavily urbanized.

homes in a pair of firestorms. The floods came three years later. December 1977 and January 1978 had been unusually wet, but without consequence. Then a huge storm on February 10 dropped 12 inches of rain on the San Gabriel Mountains in a 24-hour period, creating conditions that were an almost exact replicate of those 45 years earlier. It was one of the wettest winters in southern California since recordkeeping began more than a century earlier. Once again, there was a short and furious downpour in the middle of the night, and once again, surges of water and debris barreled down steep slopes and canyons. A new debris basin had been built at the mouth of one small canyon the year after the 1975 fire to contain just such an event, but it turned out to be inadequate.

As described by John McPhee in *The Control of Nature* (1989), at the very moment that basin filled and overflowed with debris, a pair of employees of the Los Angeles County Flood Control District were on their way to inspect the basin, approaching from the road below in a six-ton truck. They jerked the truck around and raced away from the wall of debris that came rushing toward them in the darkness. It caught up with them partway down the hill and spun the truck like a leaf on a brook, with them inside, down toward a slumbering residential street that had suddenly become a raceway for water, muck, boulders, and other objects picked up by the onrushing wave. The truck revolved and caromed off boulders and other vehicles that were being picked up and swept down the street. Miraculously the cab did not fill with water, and the truck remained upright and intact, finally crashing to a stop when it lurched against a tree and concrete block wall. Near where the truck came to a stop, the surge of debris-laden water barreled through, around, and over a house that squarely faced the lower terminus of the street-turned-sluiceway. The front of the house was buried to the eaves by a mixture of mud, rocks, and transported cars. Boulders were deposited on the roof (pl. 78). Several vehicles ended up in the swimming pool,

Plate 78. A house buried to its eaves by a giant debris flow in La Crescenta, February 1978. The objects stacked around the tree are flattened automobiles.

and a sedan was crushed to half its original height and slipped just under the eaves, like a crumpled metal letter being deposited in a mail slot. The onslaught pushed through the windows and nearly filled the house with water and debris. The people inside were floated up on a bed until they could touch the ceiling. The overflow, the wild truck ride, and the flooded house took only six minutes! Luckily the waters receded and there were no further debris flows that year. Both the occupants of the truck and the house survived. The family restored and rebuilt their half-buried home on the same site.

Threats to Chaparral

Very Frequent Fire Can Eliminate Chaparral

In a given spot, chaparral wildfires ordinarily occur at intervals of 20 years or more, as pointed out in chapter 3. However the vegetation can burn more frequently, given the proper circumstances. During summer and fall the first year or two after fire, dry and dead herbaceous vegetation from the previous spring can provide continuous fuel that will easily ignite and flash into a fire that quickly spreads between and among resprouting shrubs and shrub seedlings. Although such fires burn with far less intensity than would be the case if the fuel were mature chaparral shrubs, they can spread quite quickly and generate enough heat to kill both resprouting shrubs and small shrub seedlings. Since the immature shrubs have not produced seeds, and the soil seed bank is largely empty, populations of both perennial and annual plants can be depleted or extirpated entirely by such closely spaced fires. Under these circumstances the plant community may be shifted to other vegetation types. Where chaparral grows near the edge of another vegetation type that requires less water, such as grassland, coastal sage scrub, or desert scrub, some elements of these communities may invade and replace chaparral shrub species. Alien species accelerate this process. Sometimes the balance is shifted away from chaparral vegetation incrementally, moving further toward a mix of nonchaparral plant species each time a fire occurs close upon the heels of the previous one.

Invasive Plant Species

In some chaparral locations a regime of frequent fire disturbances encourages the establishment and growth of invasive herbaceous plants native to the Mediterranean Basin, such as bromegrasses (*Bromus* spp.), wild oats (*Avena* spp.), mustards (*Brassica* spp.), filarees (*Erodium* spp.), and star-thistles (*Centaurea* spp.). Once these weeds become well established they produce vast numbers of seeds that accumulate in the soil. As soon as winter rains begin, many of these seeds germinate and plants grow densely and quickly, their numbers making it difficult for the seedlings of native species to compete successfully with them for space and light. The establishment of invasive herbaceous weeds sometimes occurs after chaparral wildfires. The weeds displace some of the native herbaceous plants that ordinarily occur after fire. This is particularly true when chaparral grows near places where seeds can spread from established weed populations. Source locations for weed seeds include roadsides, fuel breaks, agricultural and other areas that have been cleared of native vegetation in the past such as pastures and paddocks, and other places subject to frequent human disturbance. As people spread into more and more chaparral areas, they create centers from which invasive plants can spread into chaparral following fire. These new vegetation types, often dominated by a handful of exotic weed species, are of lower stature, provide much less erosion control and capacity for water retention, contain far fewer species of native plants, and are far less valuable as habitat for native species of animals than the chaparral vegetation that they displace.

A number of species of shrubs from Mediterranean Basin ecosystems have been deliberately planted in California. Since their introduction, they have moved into the chaparral in many parts of the state, where they now are reproducing and spreading on their own. These shrubs were introduced for variety, color, slope stabilization along roadsides, and fire control. Since they evolved in a similar climate and many are tolerant of poor soils, they grow well in California without attention. Introduced shrubs now growing wild in California chaparral include rock-roses (*Cistus* spp.), rosemary (*Rosmarinus officinalis*), and three species from the legume family — Spanish broom *(Spartium junceum)* (fig. 31), Scotch broom *(Cytisus scoparius),* and French broom *(Genista monspessulana).* The brooms are classified as noxious weeds by the California Department of Food and Agriculture because they have taken over many chaparral areas, especially in disturbed habitats, and are excluding the native plant species. The tree of heaven *(Ailanthus altissima)* is becoming increasingly common in chaparral and elsewhere in California. Originally brought from Asia during the Gold Rush, it spreads and grows quickly from seeds and by root sprouting and may exceed 50 feet in height. It does well among and adjacent to chaparral shrubs, especially in disturbed areas. Because of its height, ability to spread, and tolerance of harsh conditions, it could easily displace chaparral shrubs as well as many other native vegetation types on a wide scale.

Global Climate Change

The changes in global climate predicted by the overwhelming majority of scientists will surely alter the distribution and condition of chaparral, other ecosystems in California, and the economy and life of people everywhere. It is difficult to predict with accuracy the exact nature and magnitude of these changes, but a number of general trends seems clear. During this century California is expected to become somewhat drier, much hotter in summer, and warmer in winter. A greater fraction of precipitation will probably be delivered by intense winter storms. Extreme summer heat and aridity could foster more frequent large and destructive chaparral wildfires. These fires could be followed by serious flooding if winter storms do indeed become more severe.

The combination of less precipitation and higher temperatures would favor the spread of drought-tolerant grasses, especially invasive alien species that carry frequent fires, as described above. Under this scenario a large fraction of chaparral and other shrublands across the state would be converted to grass. There could be a general shift in the range of remaining chaparral to higher elevations, moving into mountainous areas now occupied by forests and woodlands. The assemblages of populations of plants and animals in chaparral and many other ecosystems would be disrupted and rearranged, because different species would have varying abilities and opportunities to migrate to new areas quickly. The complex mosaic of vegetation types that has remained relatively stable in California for thousands of years would be disrupted. Isolated patches of chaparral now containing narrow endemic species, such as maritime chaparral in the south and serpentine chaparral in central and northern California, could easily vanish because of the lack of a nearby suitable habitat. In general a minority of species of chaparral plants and animals that are most adaptable and mobile would be expected to become more numerous and widespread, while the majority of species less able to respond to rapid change could become less common or extinct.

Options for Wise Growth

Catastrophes born out of chaparral wildfires and floods are predictable in the sense that floods from rivers and death from disease are predictable. No one can foretell exactly what is going to happen, or where and when, but by understanding the factors that underlie risk, it is possible to estimate probability and to reduce that risk through proper preparation. The balance of this section is about how to reduce the chance of loss by chaparral wildfire through informed preparation.

Many people, including myself (R.Q.), choose to live near chaparral. While writing this book a chaparral wildfire came within a block of my house, and 65 homes in my city were destroyed in a few hours (pl. 79). The motive for living in such a place is often the desire to be close to nature and enjoy the daily pleasures of observing the workings of nature. As more and more people have exercised this choice, the associated dangers have grown many times over. From the foothills of the Sierra to the narrow canyons of the Santa Monica Mountains, ex-urbanites choose to live surrounded by nature. The relentless pressure of general urban sprawl compounds this trend. From the outer ring of suburbs of the Bay Area to the scenic cities of the Central and South Coast Ranges and on to metropolitan Los Angeles and San Diego, human development of all kinds spills up and into chaparral. Given present planning practices and economic forces, this is the path of least resistance for horizontal growth. So long as the human

Plate 79. A neighborhood adjacent to chaparral, with homes that were recently destroyed by wildfire.

population of California continues to grow exponentially, as it has done for the past 150 years, the forces that have pushed development into chaparral will continue. While this spread is regulated to some degree by government planning agencies, urban growth of this kind is driven primarily by demand for inexpensive space for new homes and businesses. Some planning generally accompanies this growth so that basic needs such as roads, schools, water, and the numerous other elements of urban infrastructure are there to meet the needs of new residents. One part of the planning process is public safety, including protection from wildfires and flooding. However, this is rarely accorded as much attention as the more familiar details of roads and schools.

If the planning process were perfect all new residents near chaparral could be assured that they were reasonably safe from the hazards of wildfires and floods. These risks are only two of many factors that are considered when buildings and subdivisions are planned and approved. Well intentioned as the entire process might be, the seemingly remote possibility of a wildfire or flood is usually not the first consideration of people who propose developments, public agencies that approve the proposals, or people eager to move to a serene hillside or canyon with an unobstructed view of natural vegetation and wildlife. Moreover, older developments often do not adhere to today's building and planning standards, but there they stand. So wood-sided houses with shake roofs are tucked into canyons, chaparral and other flammable vegetation grow right to the edge of buildings, homes with unenclosed wooden decks hang out over the edge of chaparral-covered slopes, access roads are inadequate, and innumerable other hazards exist that would not be permitted in most jurisdictions under today's laws and regulations (fig. 57).

Historical circumstances and inadequate planning can be partially overcome, reducing the danger for the current residents. Property owners in established areas can be informed of hazardous conditions and be encouraged or even required to make changes. For example, when replacing a roof or

Figure 57. Chaparral wildfires move rapidly up-slope, quickly engulfing and destroying over-hanging structures too close to the edge and to the burning shrubs.

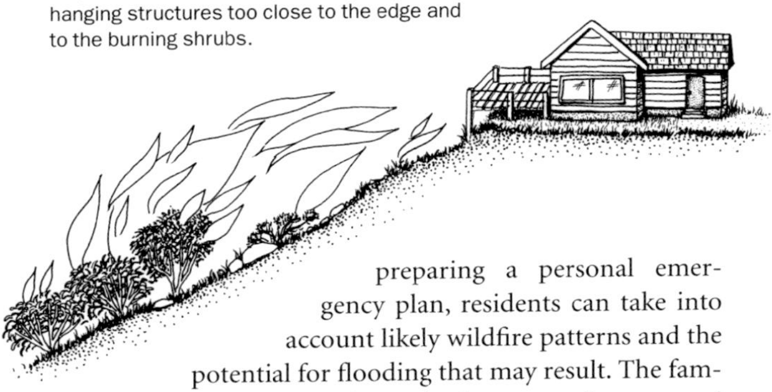

preparing a personal emergency plan, residents can take into account likely wildfire patterns and the potential for flooding that may result. The family eager to build a dream home in the mountains can, and often is required to, have a driveway suitable for emergency vehicles, have roofing material that is not readily flammable, and clear away native vegetation to specified distances from structures. After the chaparral wildfires of 2003 in southern California, written notices were mailed to 113,000 residents of San Bernardino County warning that flash floods were likely with little or no warning with each storm. These warnings were issued after a heavy rainfall, and a flash flood on Christmas afternoon had taken the lives of 16 people. People can be provided with expert advice about how to live as safely as possible with chaparral. Beyond that, however, individuals are usually free to make private decisions about how to use private property.

Individual homes that do not comply with fire-safe practices can endanger their neighbors. It is common for a single house with a flammable roof or poor clearance from natural vegetation to ignite first during a wildfire and then provide the heat and firebrands that set fire to neighboring structures.

Planning for Chaparral Safety

The larger issues that are concerned with regional public safety near chaparral are more difficult to address. These have to do with ingress and egress during large wildfires, the modification of natural vegetation, construction and maintenance of flood control facilities by public agencies using public money, and the availability and coordination of emergency workers and equipment during disasters. A chaparral landscape that is unpopulated is far easier to deal with during a wildfire or flood than the same landscape when it is studded with scattered dwellings without emergency water and accessible only by long, narrow roads.

About half of California chaparral is located on public lands, primarily in national forests, but also in other federal, state, and local government jurisdictions. When U.S. Forest Service boundaries were delineated more than a century ago, they were drawn to exclude land that was already in private hands. Consequently, the boundary lines are often quite irregular and often include islands of private land, called inholdings, inside national forests. The designation and acquisition of state lands, railroad lands, and other ownership transactions in the nineteenth century sometimes created equally fragmented boundaries between private and public land. More than a century later these lines have often become the hard, jagged boundary between chaparral and spreading urban areas. People build up against the public land, often quite intentionally, knowing that the open space beyond the backyard will remain open forever. Private inholdings are often the location of small, concentrated resort settlements that rely on the surrounding public land for recreation, water, access, and other resources and services.

This broken pattern of private-public land ownership and attendant use near cities has created a ragged and often abrupt boundary between civilization and nature. This edge is referred to as the urban-wildland interface. Such complex landscapes make it difficult to cleanly and safely separate and protect urban areas from wildfires. The neighborhood shown in pl. 79 is surrounded on three sides by expanses of chaparral, with additional flammable vegetation to the south (upper right) and between the rows of houses, which are arrayed on steep slopes and ridge tops. Weeks before this aerial photograph was taken a chaparral wildfire, driven by a severe Santa Ana wind, and firestorm approached from the east (right), burned around and through the neighborhood by following the natural and planted vegetation, and continued to the west. As is usually the case, those structures closest to the chaparral and those at the edges of steep slopes were most vulnerable. Fifteen homes were destroyed and several others damaged in a very short period of time.

Often the most practical way to increase safety at the urban-wildland interface is to reduce chaparral fuels on land that is near built up areas. Fuel breaks have been constructed for this purpose in all parts of California. Thoughtful planning of new development at the urban-wildland interface could focus on straightening out the line between chaparral and development, either by keeping structures at a uniformly safe distance from natural vegetation or perhaps by land exchanges between private and public owners. The belt of private and public land that should not have permanent structures or wildland fuels could be used for many other desirable purposes such as parks, open space, golf courses, row crop agriculture, trails for hikers, horses or bicycles, or any other use that does not entail placing flammable permanent structures close to chaparral.

The looming danger of wildfire catastrophes has steadily increased from one end of California to the other. Annual losses of lives and property from chaparral wildfires have been increasing for close to a century. The problem will continue to grow worse as long as current patterns of scattered and disorganized growth are extended ever farther into wildlands. We have ample evidence every fire season that under extreme fire conditions it is sometimes impossible to protect all lives and property from chaparral wildfires. There are several ways to reduce this danger. First, as described above, chaparral fuels can be managed in key areas. Second, settlements and structures can be made as fire safe as possible by properly designing structures and providing adequate vehicle access and emergency water sources. Third, new structures and communities can be situated in places that can be made relatively safe and can be protected by available firefighting resources. This would probably mean that new and growing towns near chaparral would have to be more compact, a recommendation that runs counter to present trends almost everywhere in California. Over the long run, better planning is the most important measure to make people residing near flammable wildlands safe.

The Public Cost of Private Decisions

Some people choose to settle in particular places with full knowledge that they have selected a spot that is at high risk of experiencing chaparral wildfires. These people sometimes argue that they have the right to assume such a personal risk. However, in the event of a fire, the safety of these individuals and their dwellings suddenly becomes a matter of public concern and expense. Residents of high-risk areas, sometimes with hazardous homes and grounds that do not follow fire-safe practices, expect the same level of emergency protection as everyone else, at public expense and often real personal risk to firefighters. When structures in dangerous places are insured against fire, they may be placed in a high-risk pool, where the full cost of the actual risk is not directly assumed by the owner. After structures are damaged or destroyed by wildfire, the victims are sometimes offered state and federal disaster relief. Assistance in the form of low-interest loans, temporary relocation expenses, and other services often supports and encourages people to more easily rebuild in the same location. These actions arise from the natural desire we all feel to help people who have had their lives disrupted by sudden and unexpected catastrophe, but it must be kept in mind that the result may be to perpetuate and subsidize settlement patterns that are inherently dangerous. It is common to require that replacement buildings be constructed with materials and in ways that are more fire safe than the originals. Over the long run, society as a whole might be better off to go a step further and entirely eliminate incentives for living in places where future fire disasters are likely. As city, county, state, and federal budgets become more and more strained and stretched, it is imprudent to allow private landholders to bill the ultimate cost of their private decision to live in a hazardous area to the rest of us. That decision is not private, and it is expensive.

One safety measure that is commonly imposed on isolated communities and homes near chaparral is to require the clearing away of native shrubs and other flammable vegetation for distances of 30 to 200 feet around all structures. This removal is important and relatively effective, but it entails replacing chaparral with other types of vegetation that are probably not native to the site. In some circumstances, where clearing is poorly planned or maintained, these places can become footholds where invasive weeds become established and spread. As more and more structures are placed in chaparral vegetation, with a cleared perimeter around each, as well as cleared strips along access roads, the carpet of continuous natural shrub cover becomes more and more tattered and disconnected. Necessary as it is, the practice of brush clearance for fire safety diminishes the ecological integrity and value of remaining disunited chaparral landscape as habitat for native species of animals, and it certainly alters the aesthetics of the increasingly unnatural landscape.

In the final analysis, Californians must decide together how much of the chaparral community we wish to retain as part of our natural heritage, recognizing that such places will always be subject to periodic fires. There is ample legal precedent for setting standards to protect and conserve the beauty that emanates from chaparral and other natural vegetation. Government everywhere already places numerous restrictions on how private property may be used in the form of zoning laws, as well as ordinances specifying aesthetic standards for many things including natural and ornamental vegetation. In many communities particular species of native trees, such as oaks, may not be harmed or disturbed on private or public property, and streamside vegetation is often protected in the interest of water quality, native fish and other aquatic organisms, and of course natural beauty.

Any seasoned firefighter can explain what precautions should be taken to reduce the probability of losing a structure to a chaparral wildfire. Most of these measures can be reduced to three categories: accessibility, defensible space, and fuel management. Firefighters arrive in fire trucks and other large emergency vehicles, often under chaotic circumstances. Street addresses should be clearly and conspicuously marked. The roads leading to the threatened structures, including private driveways, should be wide enough to permit emergency vehicles easy access. Often, emergency vehicles must reach threatened structures at the same time frightened people are attempting to depart by the same roadways. Structures near chaparral, especially houses and barns, are much safer with at least two access roads. That way if an approaching fire blocks one escape route, another can be used.

Defensible space is an area around a structure, cleared of hazardous fuels, that is large enough to place safety personnel and firefighting vehicles between the fire and the building. Hazardous fuels include buildings, attached structures, and outbuildings made of wood and other flammable materials, ornamental plants that easily burn, wood piles, automobiles, and anything else that could quickly catch fire when subjected to a sudden blast of heat, wind, and flames.

Fuel management includes a choice of building materials and architecture that makes a structure less easily ignited from the outside. The single most vulnerable point of ignition for a structure is the roof. It receives direct exposure to approaching heat and flames and is the place where burning materials carried through the air can land and start a fire. Fuel management also includes choosing ornamental vegetation of low flammability that separates, rather than connects, a structure with surrounding wildland fuels. Another very important aspect of fuel management is the chaparral itself, which should be cleared away from all structures and access ways to safe distances, equaling or exceeding those required by ordinance.

Detailed advice on planning and preparedness for wildfire emergencies is available free of charge from the California Department of Forestry and Fire Protection, the USDA Forest Service, the University of California, and various county and local governments. The Fire Safe Council provides guidance for communities. Booklets, pamphlets, brochures, on-line materials, and other resources are often free for the asking.

The Value of Chaparral

Why should we care whether the chaparral exists or not? Why should humans not simply replace the plant and animal world with the human and concrete world? If they could speak, the plants and animals could tell us what it takes for sustained life here. From studying the natural life of the chaparral we will learn lessons about careful use of water, protection of the soil by deep roots, and conservative use of resources.

By following the natural patterns of chaparral, the wasteful and often unsuccessful practice of emulating the vegetation of an English garden could be replaced by sustainable, feasible gardening methods. Rather than bright green lawns surrounded by soft-bodied, water-guzzling plants, Californians would use drought-tolerant plants to create their gardens. Xeriscape landscaping, which employs plants that grow well in dry conditions, is becoming increasingly popular, both to conserve precious water and for the natural beauty of plants native to California and other climates with dry summers. Mediterranean Europe and Australia are sources of some truly drought-tolerant plants such as rock-roses (*Cistus* spp.), rosemary *(Rosmarinus officinalis)*, oleander *(Nerium oleander)*, and a host of others. These plants are adapted to a mediterranean-type climate and thrive in Californian gardens if they are not overwatered. The chaparral's native manzanitas (*Arctostaphylos* spp.) and ceanothus (*Ceanothus* spp.), toyon *(Heteromeles arbutifolia)*, and cherries (*Prunus* spp.) also make attractive garden additions. Xeriscape landscaping is relatively easy to care for, requires far less water than traditional ornamental plants, usually requires less attention to the soil, and provides a wide variety of colors and textures. Wisely chosen plants in a carefully planned garden also create a buffer of low flammability around structures.

Chaparral shows us that it is not only possible to live with fires, droughts, and floods, but that life can flourish in spite of these "disasters." Water is always going to be a problem in a mediterranean climate. There is not enough of it to go around in California, so we need to use less and conserve more. We can appreciate that the fires that do happen are natural and life giving to the chaparral. They do produce smoke and blackened hillsides, but they also release an entire generation of wildflowers from their underground imprisonment by the shrubs above. Fires also give the shrubs a chance to renew their growth and expand their especially capable root systems. It is often best to leave these shrubs alone, not to bulldoze, bully, or change the chaparral into something else. Chaparral shrubs usually grow on poor soils and often on steep slopes, far from water, that are not well suited for other purposes. These shrubs stabilize the soil, provide lovely views, and keep the environment cooler and moister than it would ever be if cleared. There are other kinds of lessons to be learned from the animals that make chaparral their home. The efficiency of insulation and the intricate cohabitation of organisms in a wood rat (*Neotoma* spp.) nest are remarkable. Maybe within this complex is a new design for human habitations that is more efficient and ecologically sound than what we have so far devised. Clearly, we don't want to live in a pile of fuel as they do, but the nest form itself may contain some useful but as yet unrecognized principles of design. We can also appreciate that sooner or later, steep slopes and poor soils are likely to result in substantial erosion and mudslides and choose not to live in a place that is in the path of such hazards. We can use the sun to warm our homes, like the harvester ants (*Pogonomyrmex* spp.), to an even greater degree than we are already doing. Like lizards and snakes, we can power our actions with sunlight's energy, too. We can learn the intricacies of our environment so that like a Wrentit *(Chamaea fasciata)*, we find a full life within the environment we create. We need to do a better job of taking advantage of natural processes, rather than ignoring or opposing them.

For California, the future of chaparral and our future as residents here are inextricably linked. Much of the remaining privately held natural land near the growing population centers in central and southern California is chaparral and coastal sage. The rate of land development of these areas in the past decades has been, and continues to be, astonishing. Everywhere we see new housing tracts, shopping malls, and freeways being built as more and more people occupy what was once "empty" land. With this rapid spread of human settlements in California, chaparral and other natural ecological communities are in great danger of incremental destruction. The hardy plants and animals are not preadapted to bulldozers and cement. For plants and animals, sunlight, water, space, and nutrients are required for abundant life. Obviously, these simple requirements are lacking entirely under sidewalks, buildings, and highways.

Sooner or later we must strike a deliberate balance between the spreading urban and ex-urban centers of California, and the adjacent chaparral in their path. Half of the remaining chaparral is on private land and consequently available, at least in principle, for urbanization. In addition to the natural services chaparral performs free of charge, it has always been a beautiful backdrop to life in California (pl. 80).

Urbanization cannot continue to consume chaparral and other natural areas indefinitely, and few people would wish to see all of the remaining natural beauty of this exceptionally attractive state that falls outside of existing reserves consumed by development.

Chaparral has already survived many thousands of years in a demanding climate of droughts, fires, floods, heat, and cold; but the inexorable pressure of human encroachment may ultimately be a much more serious challenge. If we decide to retain as much chaparral as possible now, the plants and the animals will survive, and human Californians will benefit as well. We will continue to be surrounded by a healthy, strong, natural community that has many lessons to teach us. The vertical leaves of manzanitas may thus continue to provide for life and energy in a climate so dry most plants would perish. The roots of ceanothus, with their nitrogen-fixing bacteria, could continue to enrich the soil and beautify the landscape. And the hundreds of thousands of wild flowers, shrub seedlings, and resprouts and the myriad of animal inhabitants large and small would flourish to let us know that

Plate 80. Chaparral provides scenic hillside beauty throughout California.

life is revitalized after fire. The chaparral is ours too, and it is a rich legacy.

Chaparral does more than provide useful lessons and natural services. The ever-changing beauty of the chaparral-covered hills offers a much-needed respite from the glare of concrete and the frenzy of congested urban areas. The Hollywood Hills are famous but would hardly be improved by being entirely covered with concrete or skyscrapers. The chaparral behind and above the Hollywood sign is an essential part of its charm. The appeal of chaparral grows with familiarity. The undulating blanket of shrubs, the kaleidoscopic mosaic of flowers in spring, the carpet of wildflowers the first few years after fire, and even the pungent odors of laurel sumac *(Malosma laurina)* and other chaparral shrubs that fill entire valleys on hot summer days—all of these are parts of the natural heritage of our state that are worth saving for the future. Hillsides clothed in chaparral remind us on a daily basis of where we live, and of what it means to be Californian. The solid and durable chaparral is a reflection of the beauty, promise, and enduring spirit that continues to attract people to the state.

GLOSSARY

Adaptation The traits an organism possesses that suit it to live in its environment.

Alternate leaves Leaves that do not have a companion leaf arising from the same place on the opposite side of the stem.

Annual plant A plant that completes an entire life cycle in one growing season and then dies.

Anther The male part of a flower that contains and releases pollen.

Aquifer A geological formation that contains water.

Asexual reproduction Reproduction that does not involve union of sperm and egg cells, such as fission or budding.

Banner The upraised petal of a legume flower.

Biennial A plant that requires two growing seasons to complete its life cycle, flowering and fruiting the second year.

Burl The enlarged woody base of a chaparral shrub from which new sprouts arise after fire or mechanical damage to the stems.

Cambium (vascular) A thin layer of tissue lying just beneath the bark from which new conducting cells arise.

Carpel The female reproductive structure of a flowering plant consisting of the stigma, style, and ovary.

Cismontane Referring to the region west of the main Sierra crest.

Clone An organism formed from a cell or cells originating from another organism without sexual reproduction. A cloned organism is genetically identical to the organism from which the cell came.

Clutch A group of eggs laid all at the same time.

Coevolution The reciprocal and interdependent evolution of two or more species.

Commensal Referring to a type of interaction between two species where one organism benefits and the other is not helped or harmed.

Compound leaf A leaf with a blade that is divided into several separate leaflets.

Conifer A cone-bearing tree or shrub such as pine, juniper, or fir.

Convergence The similar appearance of unrelated organisms. The species resemble one another because they have developed similar ways of adapting to similar environments (convergent evolution).

Corm An enlarged part of an underground stem in which starch is stored.

Covey A small flock of birds, for example, quail.

Crepuscular Active at dawn and dusk.

Debris basin An excavation that traps rocks, dirt, mud, and vegetation brought down a watercourse by flooding, often having a high wall or berm at one end resembling a dam.

Dehiscence The opening of a fruit or other plant structure, allowing the escape of the seeds or other contents within.

Dissected leaf A leaf with a blade that is distinctly cut into sections that are not entirely separated from one another at their bases.

Diurnal Active during daylight.

Ecosystem A group of organisms found together that interact with one another and with the environment.

Eliaosome A nutritious structure attached to the outside of a seed coat, whose function is to attract ants to disperse seeds away from the parent plant.

Endangered A population of organisms with numbers that have declined to the point where that population is in imminent danger of extinction.

Endemic A species or subspecies that has a natural range that is restricted to a particular area.

Exoskeleton The rigid outer body covering of invertebrates including insects, arachnids, and crustaceans.

Exposure The compass direction toward which a slope is inclined.

Extirpation The extermination of a species from a particular area, but not from its entire range.

Fascicle A group of needlelike leaves that is held together at a common base. Such leaves are called *fasciculate.*

Fledging The developmental stage at which a young bird has grown enough feathers to fly.

Fuelbreak A wide, roadlike clearing in chaparral created by removing most or all of the shrubs. The purpose is to impede the progress of fires.

Gall An enlarged growth produced by a plant in response to invasion by insects, fungi, or viruses. Wasp larvae often develop within the enlargement.

Gland A bump, depression, or appendage on a plant that produces a sticky fluid, such as nectar.

Habitat The environment of an organism, the place where it is usually found.

Hard-to-wet soil Soil that contains organic chemicals that prevent water from penetrating.

Herbivore An animal that feeds upon plants.

Humus Decomposed organic matter incorporated into soil.

Hybrid Offspring of unlike parents, for example, from different species.

Inflorescence A cluster of flowers, typically in specifiic arrangements, for example, a raceme or head.

IUCN Red List A list compiled by the International Union for the Conservation of Nature comprising species that are either extinct or in danger of becoming extinct, or are declining such that they could become endangered or extinct in the foreseeable future.

Larva The worm-shaped immature stage of an insect such as a caterpillar. The plural is *larvae.*

Linear leaf Resembling a line, long and narrow with more or less parallel sides.

Longitudinal bark fissioning The braided appearance of old stems of ceanothus, produced by the overgrowth of dead parts of the stem by living parts.

Microclimate The physical conditions of a small area, often different from the general conditions of the larger area that surrounds it. This influences the presence and distribution of organisms.

Nascent inflorescence Immature flower buds destined to produce flowers in a future growing season.

Natural Diversity Data Base (NDDB) A compendium of information about the distribution and status of naturally occurring organisms in the state of California.

Nocturnal Active during the hours of darkness.

Obligate seeder A species of chaparral shrub that regenerates after fire only from seeds. Also called a *nonsprouter* or *nonresprouter*.

Opposite leaves Pairs of leaves that grow directly across from one another on opposite sides of the same stem.

Ovipositor The egg-laying organ of insects.

Perennial plant A plant that lives for several to many years, such as a shrub or tree.

Peritoneum A membrane that surrounds and supports the internal organs of the abdominal cavity of vertebrates.

Pheromone A chemical released by an animal that communicates with another member of the same species, often to attract a mate.

Photosynthesis The process of the conversion of energy from sunlight into chemical energy (carbohydrates) by plants using chlorophyll, carbon dioxide, and water.

Pinnate leaf A featherlike leaf with leaflets arising on opposite side of the midrib.

Plant family A grouping of plant species, containing smaller groups called genera, that are placed together based on genetic similarity. Each family has a particular type of flower.

Prescribed fire The deliberate burning of natural vegetation to achieve a management goal such as clearing brush.

Pupa The stage in an insect's life when the larva ceases feeding, produces a hard outer covering, and remains immobile while transforming into an adult. The plural is *pupae*.

Pupate The process of forming a pupa.

Pyrophyte endemic A species of annual plant that only grows in chaparral during the first year following fire.

Raceme A flower cluster in which the flowers are arranged along the central stalk (peduncle) on equal-length stems (pedicels).

Raptor A predatory bird that grasps prey in its talons.

Rare species A species found in extremely small numbers, often in danger of extinction.

Relict distribution A population or species with a scattered natural distribution that is the remnant of a once larger, more continuous range.

Rills Small channels made by flowing water.

Scat Animal feces cast upon the ground.

Sclerenchyma A hard, rigid type of tissue found in the leaves of many chaparral shrubs.

Sclerophyllous Referring to hard or leathery evergreen leaves that retain their shape even when dry due to the presence of thick-walled cells (sclerenchyma).

Serpentine A mineral with high levels of magnesium and other metallic elements, and low nitrogen and phosphorous. Serpentine soil is inhospitable to the growth of most plants.

Stigma The female part of a flower that receives pollen.

Stipules Small appendages at the base of a leaf, usually born in pairs.

Subshrub Low-growing plants with woody bases and soft stems that are knee to waist high.

Taxonomist One who classifies and names organisms according to their relationships with one another.

Thorax The middle section of the insect body to which all legs are attached, between the head and the abdomen.

Threatened species Those species and populations that are approaching danger of extinction.

Urban-wildland interface The boundary between developed and natural areas, for example, when houses abut chaparral.

Varve Sediment deposited in episodic layers at the bottom of a body of water.

Viable Referring to a seed capable of germinating and growing.

Warble The resting chamber made just beneath the skin of a wood rat (*Neotoma* spp.) by the larva of a parasitic botfly (*Cuterebra* spp.).

Watershed A geographical area that collects precipitation flowing out of a common drainage point.

Westerlies Prevailing winds from the west, in California blowing from the ocean across the land.

Xeriscaping Landscaping comprising drought-tolerant plants.

SUPPLEMENTAL READINGS AND REFERENCES

The supplemental readings are arranged in the order that subject matter appears in the book. Within subjects, entries are listed in alphabetical order. The references are sources of specific information quoted in the book.

General

The Control of Nature. McPhee, John. 1989. New York: Farrar, Strauss, Giroux. 272 pp. The section titled "Los Angeles against the Mountains" describes the danger and drama of floods and mud from mountainous chaparral, in the compelling prose of one of the best nature writers of our time.

The Elfin-Forest of California. Fultz, Francis M. 1927. Los Angeles: The Times-Mirror Press. 277 pp. Delightful reading, but somewhat difficult to find.

An Island Called California. 2nd ed. Bakker, Elna S. 1984. Berkeley and Los Angeles: University of California Press. 484 pp. An elegant description of California's natural history that places chaparral in a spatial context.

A Natural History of California. Schoenherr, Allan A. 1992. Berkeley and Los Angeles: University of California Press. 772 pp. An excellent ecological overview of the state, with discussions of chaparral and its organisms throughout.

Climate

Weather of Southern California. Bailey, Harry P. 1966. Berkeley and Los Angeles: University of California Press. 87 pp. A concise overview of weather phenomena in the southern half of the state.

Weather of the San Francisco Bay Region. 2nd ed. Gilliam, Harold.

2002. Berkeley and Los Angeles: University of California Press. 107 pp. An up-to-date review of Bay Area weather.

Fire

Fire in America: A Cultural History of Wildland and Rural Fire. Pyne, Stephen J. 1997. Seattle: University of Washington Press. 654 pp. A large, comprehensive, scholarly overview of the role wildfires have played in the history and culture of the United States.

Plants

Introduction to California Plant Life. Ornduff, Robert, Phyllis M. Faber, and Todd Keeler-Wolf. 2003. Berkeley and Los Angeles: University of California Press. 340 pp. An excellent overview of all California vegetation, including chaparral, for the general reader.

The Jepson Manual: Higher Plants of California. Hickman, James C. 1993. Berkeley and Los Angeles: University of California Press. 1,400 pp. The definitive volume for identifying all naturally occurring California plants. This is a technical volume intended for users with botanical training.

A Manual of California Vegetation. Sawyer, John O., and Todd Keeler-Wolf. 1995. Sacramento: California Native Plant Society. 471 pp. A detailed classification of the vegetation of California designed for land managers and other specialists. This standard reference can be easily used by nonspecialists. A Web-based version is available: http://davisherb.ucdavis.edu/cnpsActiveServer/index.html, accessed January 2006.

Plant Life in the World's Mediterranean Climates. Dallman, Peter R. 1998. Berkeley and Los Angeles: University of California Press. 257 pp. A wide-ranging overview of the vegetation of all five mediterranean climate areas, well written and illustrated.

Trees and Shrubs of California. Stuart, John D., and John O. Sawyer. 2001. Berkeley and Los Angeles: University of California Press. 467 pp. The best available book for statewide field identification of most common woody plants of chaparral.

Animals

Birds of North America: A Guide to Field Identification. Robbins, Chandler S., Bertel Bruun, and Herbert S. Zim. 2001. New York: St. Martin's Press. 360 pp. A very useful book, among many, for identifying birds in the field.

California Insects. Powell, Jerry A., and Charles L. Hogue. 1979.

Berkeley and Los Angeles: University of California Press. 388 pp. A user-friendly volume for identification of common types of insects found in chaparral.

California Mammals. 2nd ed. Jameson, E.W., and Hans J. Peeters. 2004. Berkeley and Los Angeles: University of California Press. 432 pp. Good for identifying mammals of chaparral.

Field Guide to Western Reptiles and Amphibians. 2nd ed. Stebbins, Robert C. 1998. Boston: Houghton Mifflin. 336 pp. The definitive field book for the identification of reptiles and amphibians in the western United States.

The Sibley Field Guide to Birds of Western North America. Sibley, David Allen. 2003. New York: Alfred A. Knopf. 473 pp. A guide for field identification with multiple drawings of each species and unusually comprehensive descriptions.

Living with Chaparral

California Native Plants for the Garden. Bornstein, Carol, David Fross, and Bart O'Brien. 2005. Los Olivos, Calif.: Cachuma Press. 271 pp. Useful and authoritative guide for anyone interested in growing California native plants. Beautifully illustrated.

Fire, Chaparral, and Survival in Southern California. Halsey, Richard W. 2005. San Diego: Sunbelt Publications. 188 pp. Ways to deal with chaparral wildfires in southern California, with sections written by a number of experts. This is a good source book for preparing for wildfires.

References

California Department of Forestry and Fire Protection. 2002. *Fire Hazard Zoning Field Guide.* http://osfm.fire.ca.gov/zoning.html, accessed January 2006.

Clar, C. Raymond. 1959. *California Government and Forestry.* Sacramento: California Division of Forestry.

Dana, Richard Henry. 1949. *Two Years before the Mast.* Garden City, N.Y.: Doubleday.

Dole, Jim W., and Betty B. Rose. 1996. *An Amateur Botanist's Identification Manual for the Shrubs and Trees of the Southern California Coastal Region and Mountains.* North Hills, Calif.: Foot-loose Press.

Keeley, Sterling C., ed. 1989. *The California Chaparral: Paradigms Reexamined.* Los Angeles: Natural History Museum of Los Angeles County.

Robinson, Alfred. 1925. *Life in California before the Conquest.* Thomas C. Russell, ed. San Francisco: Thomas C. Russell.

Vancouver, George. 1984. *George Vancouver: A Voyage of Discovery to the North Pacific Ocean and Round the World, 1791–1795,* vol. 3. W. Kaye Lamb, ed. London: The Hakluyt Society.

ART CREDITS

Photographs were generously provided by a number of colleagues. Individual slide credits are given below; those not listed are from the authors' personal collections.

Plates

STEPHEN DAVIS, Pepperdine University 6, 7, 10, 29, 33–36, 39, 44, 49, 51, 60, 74

JACQUES DESCLOITRES, courtesy of NASA 12

WILLIAM FOLLETTE 59

ROBERT GUSTAFSON, Los Angeles Country Museum of Natural History 4, 41, 42

LINDA HARDIE-SCOTT, The Nature Conservancy 1, 42, 52, 54, 57, 58, 71

CHARLES HOGUE, Los Angeles County Museum of Natural History 69, 70, 72, 73

FRANK HOVORE/DEDE GILMAN, Hovore Associates 5, 32, 67, 80

GLENN KEATOR 16

PHILIP RUNDEL, University of California at Los Angeles 40

TIMOTHY THOMAS, U.S. Fish and Wildlife Service 3, 25, 61, 68

SHERRY WOOD, Gonzaga University 26

PAUL ZEDLER, University of Wisconsin 13, 32, 56

Figures

6 Chamise shrubs drawn after Arthur W. Sampson, "Plant succession on burned chaparral lands in northern California," *Calif. Agric. Exper. Stn. Bull.* 685, 1944.

7, 8, 10, 12–14, 16, 18, 19, 33   Drawn in part from Hazel Gordon and Thomas C. White, Ecological Guide to Southern California Chaparral Plant Series: Transverse and Peninsular Ranges: Angeles, Cleveland, and San Bernardino National Forests, RS-ECOL-TP-005 (San Francisco: Cleveland National Forest, U.S.D.A. Forest Service, 1994).

25   Cone redrawn with permission from Jim W. Dole and Betty B. Rose, *Shrubs and Trees of the Southern California Coastal Region* (North Hills, Calif.: Foot-loose Press, 1996).

52   Whiptail drawn after photograph by Nathan W. Cohen in Alden H. Miller and Robert C. Stebbins, *The Lives of Desert Animals in Joshua Tree National Monument* (Berkeley and Los Angeles: University of California Press, 1964).

57   Drawn from figure 22 of Klaus W. H. Radtke, *Living More Safely in the Chaparral-Urban Interface,* General Technical Report PSW-67 (Berkeley, Calif.: Pacific Southwest Forest and Range Experiment Station, U.S. Department of Agriculture, 1983).

INDEX

ABOUT THE AUTHORS

Sterling C. Keeley is Professor of Botany at the University of Hawaii in Honolulu. She served as the scientific adviser for the Chaparral Hall at the Los Angeles County Museum of Natural History and edited *The California Chaparral: Paradigms Re-examined* (1989). In California, her research focused on the fire-following plants of the chaparral.

Ronald D. Quinn is Professor of Biological Sciences at California State Polytechnic University, Pomona. He has published widely on effects of fire and herbivory on chaparral, and other mediterranean ecosystems of the world.

Marianne D. Wallace has been a natural science illustrator and educator for over 30 years. She lives in the foothills of southern California, sharing her toyon-bordered backyard with wood rats, Mule Deer, Wrentits, and other chaparral species.

Series Design:	Barbara Jellow
Design Enhancements:	Beth Hansen
Design Development:	Jane Tenenbaum
Composition:	Jane Rundell
Indexer:	Thérèse Shere
Text:	9.5/12 Minion
Display:	ITC Franklin Gothic Book and Demi
Printer and binder:	Golden Cup Printing Company Limited

Introduction to California Desert Wildflowers, Revised Edition, by Philip A. Munz, edited by Diane L. Renshaw and Phyllis M. Faber

Introduction to California Plant Life, Revised Edition, by Robert Ornduff, Phyllis M. Faber, and Todd Keeler-Wolf

Introduction to California Chaparral, by Ronald D. Quinn and Sterling C. Keeley, with line drawings by Marianne Wallace

Introduction to the Plant Life of Southern California: Coast to Foothills, by Philip W. Rundel and Robert Gustafson

Introduction to Horned Lizards of North America, by Wade C. Sherbrooke

Introduction to the California Condor, by Noel F. R. Snyder and Helen A. Snyder

Regional Guides

Sierra Nevada Natural History, Revised Edition, by Tracy I. Storer, Robert L. Usinger, and David Lukas